Animals

Karl König, 1962 (Photo: Carlo Pietzner)

Animals

An Imaginative Zoology

Karl König

Floris
Books

Karl König Archive Publication, Vol. 13
Subject: Agriculture, science

Karl König's collected works are issued by
the Karl König Archive
in co-operation with the Ita Wegman Institute
for Basic Research into Anthroposophy

Translated by Richard Aylward
Series editor: Richard Steel

First published in German as essays in *Die Drei*
between 1956 and 1966
First published in volume form as *Bruder Tier*
by Verlag Freies Geistesleben in 1967
First published in English as three volumes *Penguins;*
Swans and Storks; and *Elephants* in 1984, 1987 and 1992
Full publication details are shown at the back of the book
This revised edition published by Floris Books in 2013
Second printing 2022

British Library CIP Data available
ISBN 978-086315-966-4

Contents

Introduction

The unique aspect of this work, *Animals* – a literal translation of the original German title would be 'Brother Animal' – is that Karl König not only collects together evidence from the visible world around us and gives it consideration from the usual descriptive scientific viewpoint, but that he applies heart-born knowledge, an asset yet young and in its dawning era for humankind. This fruit of true Goethean spirit lived in König's way of observation, judgment and description, and the light of this particular wisdom rays into the animal kingdom in the chapters of this volume. Symptomatic of this specific mode of heart knowledge is the process of ripening that is entailed in the development of a theme. Karl König wrote these essays during the last ten years of his life as a fruit of life-long striving, the roots of which reaching well back into his youth. Even at that time he was able to consciously pinpoint the task ahead: to connect scientific exactness with spirit knowledge and religious devotion. The character of his individual path is revealed when the spiritual undercurrent of his approach and the intensive powers of empathy accompanying his studies of natural science are considered.

Karl König possessed a strong social conscience that already showed itself in his childhood, and a tireless will to heal which expressed itself throughout his life as physician and curative educator. Although both of these facets were directed at the human being in particular, his attention always extended to a holistic approach, and being aware of the necessity to heal the entirety of nature. We are reminded of St Paul's words about the redemption of creation (Rom.8:19–22). And within this his love of animals and a sense of brotherly connection to them never left him. His mother noted a small episode in her memoirs in 1966, which illustrates this quality:

> Once my son did not return home the whole night; we assumed he was staying with the Bergels. Early in the morning he came in looking bedraggled and dirty.
> 'Well, where have you been?'
> And he replied, 'With the police.'
> And for what reason? He had given a speech in the city's park about how badly animals were treated. The city department just catches all stray dogs and throws them onto a trailer in the most cruel fashion ... their howling was so pitiful. And so my son gave a speech about it in the park. A huge crowd gathered round him, but that was forbidden in Vienna so a policeman came and arrested him; he wasn't even allowed to phone us. That's the difference between the Austrian police state and Britain![1]

Karl König was born 1902 in Vienna. He gradually began to distance himself from his Jewish background from the age of twelve. He turned to Christianity, at first to the Catholic faith, but then soon realised that the doctrinal confines were not his way. His religious inwardness was already deeply connected to the earnest striving for philosophical and scientifically attainable truth and a strong quest for social renewal towards real brotherliness. König experienced the peak of his youthful striving at the age of 18, around the time of his first moon node,[2] on May 4, 1921:

In recent days I underwent an experience hitherto
unknown to me. The physical environment disappeared
from around me for hours and I was able to glimpse
the most inner of worlds. I understood that we are part
of the whole of eternity, the universal All; and that the
I *is* the universal All. In these hours I experienced my
consciousness as far greater and more far-reaching than
the stars. Time seemed to stretch out for very, very
long periods. It seemed to me as if in these hours I had
grasped 'thinking' for the very first time.[3]

With this, König described the inner portal he entered on his
path towards studying at the university of Vienna. Because of the
type of school he had visited he first had to study Latin for a year;
during this time he took courses in botany, zoology and biology,
as well as physics, chemistry and higher mathematics. Next to
this he still kept up his studies of poetry and music that he had
begun in his youth. He particularly loved Beethoven, Bruckner
and Mahler, and also liked to play the piano, so that music was a
central part of his education and he even considered a career as
conductor. But in the end he was led to study medicine by his
profound interest in the human being and the decisiveness of his
will to heal. From the beginning of his studies, he never saw the
human being in an isolated situation but always in the context of
his social environment.

At the time when Karl König started at university he wrote
in his diary that the whole world of scientific research began to
disclose itself to him with 'excessive might', yet he prophesied
with equanimity:

The sea of materialism will rush in upon me, but I shall
stand fast. The world and the universe are full of God
and full of angels and wonders, full of goodness and
anger and full of will.[4]

At the same time as his first experience with viewpoints of materialistic mathematical research necessary for medical studies, he also came across Goethe's scientific work, whereas he had previously only known and respected him as a poet. Thus König began to grasp the spirit of the true Being of Life in the context of natural science that he had already taken hold of within his own thinking:

> [Goethe's] botanical and anthropological-morphological descriptions were like redemption for me. I felt directly addressed: these were the gateways that would lead to possible answers. In the Goethean conception of nature I met something that enlivened my thinking in the same way that the New Testament had awakened my senses for a new dimension of existence. Now, for the first time, the study of anatomy, embryology and histology could become a daily source of the most profound and heartfelt joy. Bones and muscles revealed new worlds to me. The idea of metamorphosis gripped my attention deeply, and through this I came to know the working of creative formative forces in nature. I also began to grasp the identity – the absolute 'oneness' – that exists between these creative forces and our thoughts. Outside in nature these formative forces work in such a way that they bring all organic forms into being, while inside, in the human soul, they are the creators of our thoughts and ideas.[5]

In this way König's own world view formed itself between a spiritually imbued cosmos and the natural life of plants and animals on earth. In all this, his focus was always the question of human development as seen in embryology and evolution. The deep questions he had especially at that time remind us of a similar point of destiny exactly a hundred years earlier in 1827 with Charles Darwin whose questions were very similar. Darwin, however, decided to forego all artistic activities that had previously played such an important role hitherto, in order to be able to dedicate himself completely to research into natural science. This he came to regret later in life.[6]

During his studies König also worked for three years at the Embryological Institute. In his diary we find this entry:

> I learned the methods of exact research and taught the art of using a microscope to hundreds of students. Most significant of all, however, I became thoroughly familiar with human embryo formation and development.[7]

The main phylogenetic and ontogenetic law drew his attention and made him aware of the spiritual aspects of human and world evolution. In the appearance of present-day animal forms he could see early steps of development taken in the context of human evolution, whereby the animals still carry features of a bygone stage of mutual achievement, as brothers instead of ancestors to humankind. Out of the spirit of this viewpoint König began his animal studies, the last of which he wrote just before his death. He gave that one the title *Bruder Pferd* (brother horse), leading to the use of the title *Bruder Tier* for the book published after his death. It is a term indicating a principle that needs to be regained today on a new level of consciousness; a principle that König worked with in many ways throughout his life. In the ancient mysteries one experienced the animal in a special way as standing between the plant kingdom and the human being. Rudolf Steiner described this briefly:

> Whoever finds their way into the deeper levels of the circumstances comes to realise the plant in an inverse human being. It has its roots downwards, and its stem, leaves, stamen and pistil reach upwards; the pistil constitutes the female and the stamen the masculine organs of pollination. With naive innocence the plant opens its organs of pollination towards the sun because the sun is the stimulator of pollination. In reality the root is the head of the plant which stretches out its organs

of pollination into the widths of space and whose head is drawn in towards the centre of the earth. With the human being this is opposite: his head is upwards and the organs which the plant stretches out towards the sun are downwards. The animal is in the middle with its body in the horizontal. A half-turn of the plant brings it to the position of the animal, a full turn that of the human.

This was expressed by ancient mystery wisdom with the age-old symbol of the cross, adding in the way that Plato expressed it, 'The world's soul has been fixed to the cross of the world's body,' that is, the soul of the world is present everywhere, but it has to work its way upwards through these three stages; it has to journey on the cross of the body of the world.[8]

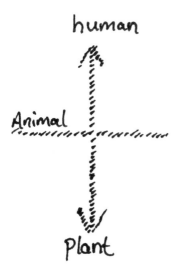

The earth cross of the world soul.[9]

Rudolf Steiner spoke about this context of the realms of nature frequently, particularly around 1905/06. One example that can be of interest in connection to Karl König is to be found in the cycle of lectures, *An Esoteric Cosmology,* given in Paris:

Suppose there are two brothers, one of whom is handsome and intelligent, the other ugly and dull. Both proceed from the same father. What should we think of those who believed that the intelligent brother descends from the idiot? This is the kind of error Darwinism makes in regard to the races. Human and animal have a common origin; animals represent a degeneration of a single common ancestor, whose higher development is expressed in the human being. This should be a source of pride, since the higher races have been able to develop only because of the lower kingdoms.[10]

Directly after this Rudolf Steiner continued by mentioning the humility of Christ, washing the feet of the disciples (John 13). This is the gesture we can experience in the mood of Karl König's writings about animals and which expressed itself in the term 'brother horse' in the last essay he wrote.

Karl König described the animal often as a creature imbued with cosmic ether substance. This is close to what Goethe found through his work on the theory of metamorphosis, describing life forces that manifest in 'typical' characteristics, or archetypes, progressing upwards through three stages from the plant form *(Urpflanze)* through the animal *(Urtier)* to the human form. Goethe expressed in his diary of 1790:

I became utterly convinced that one can indeed observe a general archetype that threads its way forwards by means of metamorphosis through all of creation, manifesting in its various parts as certain mediate stages, and that this archetype is still recognisable at its highest level, in humanity, even though it there recedes into covert diffidence.[11]

This defines the innermost kernel of Goetheanism in natural science that König was able to discern as a young student, seeking to further advance it during the rest of his life. The extent of his interest in zoology is even reflected by the selection of books in his library, around 200 of them pertaining to animals, many with his hand-written notes in the margins.

However, the foremost task in Karl König's life was not his written work, but in founding and developing the Camphill Movement, which began in Scotland during the Second World War and has since spread across the world. From the beginning, the main focus was on children, youngsters and adults with special needs, striving for community forms that can provide a basis for healing and therapeutic practice. At the same time König was striving for social reform out of new spirituality. He began this task together with young friends from Vienna who had been members of his youth group, learning about anthroposophy from him.

Amongst König's many friends and colleagues of that time, Eugen Kolisko must have a special mention; they met in their youth in Vienna. In 1920 at Rudolf Steiner's suggestion Emil Molt asked Kolisko to become the physician at the first Waldorf School in Stuttgart. Steiner had given Kolisko important insights about cosmic principles for a classification of the animal kingdom, which Kolisko researched. But it was not only medicine and zoology that connected König and Kolisko; they had many common interests, like agriculture, nutrition and social questions, and often worked together at conferences during the 1920s and 30s. They met again as refugees in Britain in 1939. Hopes of further collaboration were, however, prematurely ended by Kolisko's death in November 1939.

In 1930 Kolisko had published *The Twelve Groups of Animals* showing their connection to the zodiac, which was directly significant for König's work leading to the essays presented here. Although König had only just arrived in Scotland and was occupied with new community at Kirkton House (the forerunner of Camphill), he wrote three booklets on the same theme in 1939–40.[12] His preparatory notes ran to about a hundred pages in his notebook; he must have written them during his weekly train journeys to Dundee where he prepared for his British medical degree because his qualification from Vienna was not recognised. In spite of the many adverse circumstances it was important for him to finish this zoological work, which he later deepened and differentiated in many lectures, courses and seminars, right up to a medical seminar for physicians and eurythmists in 1963.[13] Together with this present volume König's writings on animals is a wonderful example of the way he worked: founded on a profound and wide knowledge of the subject area, he was able to penetrate the theme in an imaginative fashion that is only possible out of the forces of thinking awakening within the heart.[14] When König was a seven-year-old Rudolf Steiner had spoken prophetic words in the very city of Vienna: 'Other abilities like thinking through one's heart will develop within the human soul during its transformation into the future.'[15]

König was thus able to open a pathway for many people into greater depths of experience, in this case for the animal kingdom. His approach can lead the way for us all, but in this case he is especially a guide for anyone teaching at any level. From his contemporaries we have heard many descriptions of how his imaginative mood could include listeners at his lectures, inspiring them and inciting images for them; but especially when he spoke about animals, as he often liked to do, a simple gesture and a few paces through the auditorium sufficed, and one felt the very presence of an elephant, or whatever else.

Karl König had worked on the final essay, 'Brother Horse,' during his last days; page 48 of his manuscript only had the subtitle 'Epilogue'! Thanks to the sensitive connection Fritz Götte, the editor of first German edition, had to König we can sense a little of what König might have written. Using König's extensive notes Götte rounded off the book himself. In the foreword to that edition he tells about the conversation he had had with Karl König

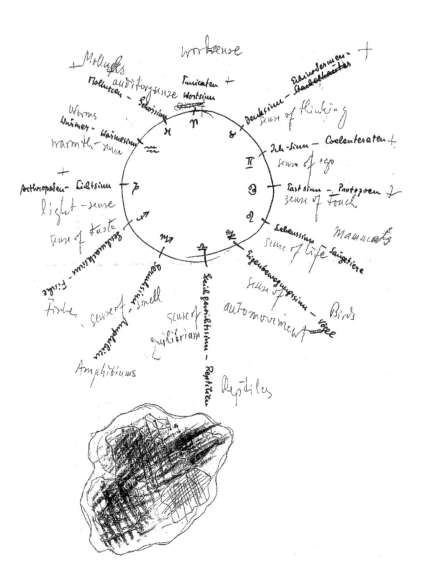

A page from Karl König's notebook

in 1959 where they had come to the decision that König should continue the series he had just begun:

> Then the idea came up that we should try to save more and more animals from their endangerment or even extinction due to human behaviour, by building a Noah's Ark within the human soul.

This remark, that was made three years before Rachel Carson's ground-breaking book, *Silent Spring,* was published, has certainly lost nothing of its urgency and relevance.

Imanuel Klotz

Editor's note

For this edition I have completed the translation by adding passages which previously were shortened. Crispian Villeneuve has added a translation of the final epilogue, which was not included in the previous edition.

I would like to thank Professor Wolfgang Schad and Dr Johannes F. Brakl for thoroughly looking through the text and providing updated zoological research results. In a few places these were simply corrections that have been incorporated within the text; in most cases they have been added as endnotes.

Finally, thanks also go to Graham Calderwood who has made new drawings to illustrate the texts. Graham worked with Karl König in Camphill in Scotland and also illustrated his 1963 lectures *The Animals and their Destiny*.

1. The Origin of Seals

The migration of animals

A constant urge to migrate pervades the entire animal kingdom. There is hardly a family or a species for which migrating and returning are not an integral part of existence. The journeys may be long or short. Certain species cross oceans; others fly over entire continents.

This migration occurs in the most varied forms. It can, as with birds, extend over thousands of miles; some kinds of butterflies cross high mountain chains. Reindeer migrate over large stretches of the north. Eels, which breed in the Sargasso Sea, move eastwards and go up the rivers of the Eurasian continent. Salmon take the reverse path, from streams and rivers back into the ocean. Migratory locusts, which destroy everything in their path, and the hordes of migratory ants, swarm and creep in overwhelming numbers across vast land areas.

The migration of herring, the emergence of sturgeons, seals, sea lions and penguins at definite times of the year, and their disappearance after shorter or longer periods, are all partial manifestations of this incessant coming and going.

Is there a single cause at the root of this roaming and migrating? It is an uncommonly complex phenomenon, which appears to depend on the most varied conditions. Each species has its own form of migration, and this is just as characteristic for it as the structure of its body or the arrangement of its teeth. Some migration is subject to seasonal rhythms; some follows the phases of the moon. Often the mating and birth periods are closely connected with the change of place. There are also nomadic animals which follow their source of food, and others that are suddenly seized by a migration mania and, like the Scandinavian lemmings, run straight to their deaths. When we try to find the essential features of migration, amid this multiplicity of phenomena, we find that the only way is to define the idea of migration as widely as possible. The more comprehensively we learn to look at the phenomenon, the more clearly its essential characteristics come to light.

A beehive in which regular activity has reigned for weeks on end, which has gathered nectar and pollen, cared for the larvae, taught the young bees their occupations and functions, is suddenly seized by a general unrest. All regularity is interrupted. Foraging has already been low for several days; the queen cells, in which future queens are almost ready to come out, are strictly guarded. Then, often when the sun breaks through after a short period of rain, in a moment the bees begin to swarm. The old queen and a large number of young bees leave the hive and go as a compact swarm in search of a new hive.

The Norwegian lemming, a kind of vole, can live for years in the plateaux and high moors of the northern mountains – solitary, hardly noticed, very shy and withdrawn. Then one summer, when breeding has been more plentiful than usual and a multitude of young lemmings populates the heathlands, the urge to migrate breaks over them. By the thousands they mob together, become quarrelsome and aggressive, run through woods and bush, cross rivers, plunge into deep gorges, suffocate one another, pressed body to body, and always they rush on towards the west until they reach the coast, and at last, driven on without pause, they perish in the ocean.

Penguins, having avoided the shores of Antarctica for months, suddenly emerge, as if touched with a magic wand; by the hundreds they come out of the ocean, and, crowded nose to tail, populate the solid land. Here they build their enclosures – pits or small depressions surrounded by stones. In these they lay their eggs and hatch their young. When the youngsters have learnt to swim, the penguins troop back into the sea and disappear, no-one knows where, for the rest of the year.

These examples could be supplemented with many others. In each case we see that, along with the occurrence of migration, another element appears. The individual creatures band together into greater or smaller groups. Birds, fish, insects – they all swarm together and set off towards their goal in serried companies.

Many explanations for this behaviour have been offered. Each one of these theories contains a bit of truth, but none does justice to the phenomenon itself in all its aspects. Certainly hunger, sexual instinct, and expectation of death, play a role in it; but what brings about this banding together? What impels individuals to seek common traits with members of its species, and only with them, and to undertake these migrations, journeys and wedding flights? Why do thousands of penguins, tens of thousands of seals, millions of herrings, eels, and sardines suddenly crowd together? Some kinds journey together, others settle down in numberless multitudes in certain places.

Can we grasp intuitively what occurs in these moments of animal life? Of course, one can try to put the responsibility for it on a glandular function, the sudden awakening of an instinct, and other things; but the glands change their functioning, the instincts awake, because something higher, stronger, permeates the whole species and transforms it.

What happens to a group of birds in the autumn when they take up the wanderer's staff, so to speak, and fly southwards? A sudden unrest comes over them; banding together occurs and the travelling begins. Birds of passage which are kept in cages experience the same unrest when their non-captive relatives are beginning to migrate. Lucanus writes:

> The frenzied captive flaps about tirelessly in its cage, and often damages its plumage almost beyond recognition ...

This proves with certainty that it is not outer causes
which impel the migratory bird to its travels, but that if
follows an all-powerful drive which masters it completely
and cannot voluntarily be suppressed or changed. The
bird of passage migrates because it must![1]

But why must birds migrate? Because all living creatures,
including human beings, are imbued with certain rhythms of
life. It is not justified to compare even approximately the migratory drive of animals with the human being's desire for travel
and search for knowledge. This error has repeatedly blocked the
way to true insights. Animals and birds migrate and return in the
same way that human beings sleep and wake.

Birds preparing for their flight south experience a change of
consciousness to which they must yield. It is an experience of falling asleep, of evening, which comes over them. Then they begin to
dream of the south, and each species has its communal dream; its
members come together in the experience of this dream and find
their way, like sleepwalkers, to the land of their dreams. A change
of behaviour comes about in all of them. Lucanus relates:

On the Courland Spit I could often observe how
migratory falcons and sparrow-hawks journeyed in the
immediate vicinity of thrushes, starlings, finches, or
other small birds, without showing the least desire for
prey, and all the small birds paid no attention whatever
to the otherwise dreaded predators, but continued their
air journeys quite near them, not even changing their
direction of flight in the least degree.

This can be explained only by the fact that all these birds, predators and prey, have become dreamers. A light sleep has come over
them, and during their stay in the south they will not wake from
this sleep until the early morning of their return journey dawns.
Then they begin to strive back towards their homeland, back to

daytime work – nest-building, hatching, bringing up the young. When this work is done, the evening of departure, the dream of the south, begins to come over them once more.

Here the powerful background which underlies all animal migration is revealed. As we human beings sleep and wake in the rhythm of the daily rotation of the earth, so are birds and animals subject to a similar, but yearly, rhythm. Not the earth, but the interplay of earth, sun, and moon sets the rhythm of their sleeping and waking. Migrating and returning is an experience of falling asleep and awakening for the group souls of the individual species. The overwhelming power of this phenomenon is simply not to be explained on the basis of instincts, drives and modes of behaviour alone. A mighty soul-breath passes through the individual species: breathing out, it lifts them from their daytime work into a dreamland; breathing in, it leads them back to everyday life.

The life of the seals can only be thoroughly understood when we see these great breaths which flow across the earth. This large animal group, about which so many mysteries are woven, is governed by this rhythm in a special way, as we shall now show.

The annual cycle

The life of most seals is determined largely by an alternation between migration and rest. The swinging of this life-pendulum is emphasised by the fact that one period of it takes place mostly in the element of water, and the other entirely on dry land. The length of the periods varies with the species and their environments. Some spend half the time on land, others only a few weeks.

In all species of seals the young are born only on land, never in the water. Pairing as well takes place on land, soon after the birth of their offspring; and the pups, helpless and utterly dependent on maternal care, come to know the sea only after some time, under their mothers' guidance and leadership. In small pools on the shore they receive proper swimming instruction, until they have mastered the element of water. Then it is again time for the wide spaces of the ocean, and when they return to dry land, they are grown up.

The young seals grow and develop in swift steps. The milk teeth, in those species which have them, are lost before, or soon after, birth. The weight increase in seal pups is about one and a half kilograms (3 lb) a day. Thus the young grow amazingly fast, and are weaned just a month after birth.[2]

Usually only one pup is born at a time, and if the mother, forgetting herself, swims out for a day or two and does not look after her child, or perhaps does not return, it will starve. The young begin to whine pitifully, and real tears run out of their great dark eyes.

After the suckling period, the pups stay for one more month, unwatched, on shore. The mothers have lost interest in them and live in the harems of their mates. The young continue to grow, although taking hardly any nourishment; their coat changes colour, and when the storms of the oncoming winter begin and cold days set in, all the seals, young and old, move out into the ocean. Where they go is not always clear, but they undertake long journeys. Seals ringed in Norway have been found the following year in southern Sweden, Scotland, and Iceland. The majority go back to their old breeding grounds, and their return proceeds according to strict rules.

The ursine seals living in the northern Pacific, from Alaska to Kamchatka, begin to appear at their breeding grounds towards the end of May. First come the older, powerful males, soon followed by the younger seals. Throughout the month of June there is a continual struggle for suitable nesting sites. Each of the older bulls encircles his domain – a few square metres in size – with a few stones and clods of earth, but above all with his anger and jealousy. Thousands of these precincts lie next to each other; and the young males, who cannot yet claim a nesting place, stay more or less respectfully in the surrounding area.

Then, around midsummer, and on into the first days of July, thousands of females climb out of the sea and allow the bulls to lead them into their domains. The stronger the male the greater the number of his wives. Not many days after this the children are born, are suckled and raised, and at the same time new matings take place.

In the Antarctic, where the southern seals live, the same process occurs, only it begins in November and lasts until March of the next year.

During all this time the seals take in no food. Their life is leisure and idleness now, not hunting and preying. It is also strife among the males; it is love-play and comfort. The little pups grow and get up to their childish mischief. All observers, such as Lockley, who have lived at such a breeding ground for many weeks, have always been captivated by the magic world of this existence.

What goes on at all these places where seals settle down on land is a picture of their involvement in the course of the sun. When the sun is climbing towards its yearly culmination, the seals climb out of the sea on to land. They leave the sea not because the climate is getting warmer and living conditions are better on land. The sun carries them out of the depths of the water into the heights of the air. It is a summer lulling-to-sleep that takes place. The seals are permeated by dream-pictures, and must give themselves to them.

This is the reverse of what happens with migratory birds. For them, breeding and hatching are day-work. The seals have moved it into their realm of summer sleep and dreams. Such differences are of great importance for the study of biology and the history of the earth, and deserve to be investigated much more deeply than they have been so far. When autumn comes and the sun loses its strength and sinks downwards, the seals wake up again. Autumn is their morning. Then they go back into the water and become predators and hunters; now they begin their day-work.

This periodicity within the year has yet another aspect. Within the great circle of mammals, the seals *(Pinnipedia)* form their own order. Some researchers class them as predators; certain features are indicative of dogs. Their character and way of life make it difficult to include them in any other order. They are also capable of amazing swimming feats; they are swift, bold, and eager to attack. In the sea they do not seem to live in pods but remain solitary, only loosely connected with other members of their species. Their movements are characterised by exceptional skill and suppleness.

On land, however, seals are clumsy; since the upper arms and thighs have remained, in a very shortened form, within the skin and the rest of the limbs have been transformed into fin-like appendages, locomotion is greatly hindered. They crawl, propping and dragging themselves along the ground. They also give up their predatory ways. They draw together in small pods reminiscent of hoofed animal herds. One bull rules the pod, which consists of a varying number of females and their young.[3] Elephants have similar social tendencies.

Thus the seals swing back and forth not only between water and land, but also, in their character, between predator and hoofed animal. During their waking period they resemble the former; during their dreaming period, the latter.

In addition, seals have certain traits that are almost human, or at least seemingly anthropoid. A mother gives birth to only a single pup or, rarely, she may have twins. The young pup can whine and cry tears, and even has milk teeth. The human expression of the seal's face comes about because the eyes are large and round, and because the head has an almost spherical top. In some species, especially the common harbour seal, the forehead projects over the eyes, and since the snout is not pushed forward too far, the resemblance to a human face arises.

Once I was standing at dusk on the beach at Tintagel, quite near to Merlin's cave, and the sea was singing its dark song, when a seal suddenly emerged out of the water. He looked at me with questioning curiosity, and our eyes met. It was a look such as I have scarcely ever exchanged in such immediacy with an animal:

a look without fear, without shyness, with full understanding of the situation. It was then I began to awaken to the riddle of these strange creatures.

Habitat and origin

In the history of the earth, no ancestors of the seals have been found.[4] The skeletons and skeletal impressions described by paleontology show the same structures found in the species living in our time: the stumpy limbs, the regressive tail, and the characteristic formation of the teeth. The findings have been made almost exclusively in those geological layers corresponding to the two initial periods of the Tertiary – the Miocene and the Pliocene.

The facts indicate clearly that the order of seals arose comparatively late, and probably quite suddenly. Wherever their remains have been found, they show the characteristic features, with no earlier or intermediate stages. Suddenly, in full and perfected form, they are there.

At first one can hardly doubt that the seals were originally land animals. They are still lung-breathing, and their newborn cannot live in the sea during the beginning of their life. For the moment we must suppose that all seals went from land into the water, and, as in dreaming remembrance, return each year to the home of their origin, guided by the sun.

Where are the shores to which they find their way, and where do the seals have their chief breeding grounds? The latest surveys show unequivocally that the original centres around which they lived were the two polar areas. The Arctic as well as the Antarctic are still their habitat today.[5]

Certain species, such as leopard seals, certain kinds of sea lions and elephant seals, inhabit the Antarctic. Others emerge to mate and raise their young on many islands and peninsulas extending round the Arctic: Greenland and Iceland, the eastern and western shores of northern Canada, and the islands stretching between America and Asia and reaching from Alaska to Kamchatka and Sakhalin. Since seals pass their sleep and dream of summer without taking in food, the barren, stony, often icy and snowbound world of the polar shores is a possible habitat for them.

Some seals, however, especially among the common seals, find their way along the coasts towards the south. In Europe they are regularly seen in Ireland, Wales and Cornwall. They can appear in Portugal; and a certain group, the monk seal, even inhabits the shores of the Mediterranean. The further south they go, the less clear the yearly periodicity of their existence becomes. They play back and forth between water and land and lose their original life-rhythm.

They also inhabit inland seas, such as Lake Baikal and the Caspian Sea, and this could be a key to their geographical distribution. Perhaps the waves of the ice-age floods, pushing southwards, first brought them there, and left them when the floods retreated.[6] The paleontological findings also support this supposition: seal skeletons have been found in southern Russia, in Hungary, in Italy, and even in Egypt.

It is from the north, around the Arctic, that the eared seal seems to come. There, near the Pole, was their original home, and many still return there today. The southward-moving glaciers of the ice-ages brought the seals with them. The wider the ice-belt became, the further southwards it pushed the original polar animals. From the Antarctic as well, individual branches of the seal family push northwards along the coasts of South America, reaching as far as Patagonia; and the sea lions extend beyond the equator. South Africa, Australia and New Zealand are likewise habitats of the seals.

Can we form a coherent picture of the seals out of these few paleontological and geographical findings, and with it decipher their background within the history of the earth? From the Arctic regions they have been driven step by step into the temperate zones, on successive waves of the ice-ages, but the polar regions have remained their home. Now the nearer we come to the poles, the more the sun's yearly course dominates over the daily rhythm, so natural for other parts of the earth. Earth day and earth night are transformed into polar day and polar night. For months on end the sun does not appear over the horizon, until it rises in the spring, not to set again for many weeks. Here we meet the same periodicity that is inscribed into the life-rhythm of the seals. During the time of polar night, the sinking sun takes its seal-children with it, and they plunge into the waters of the great oceans. When the sun rises for the polar day, all over the earth the seals follow it and clamber onto the land. The whole order of seals is permeated by this polar sun-rhythm. And it is this fact that first enables us to understand the geographical distribution and life-style of these animals.

The prehistory of seals – which we have been able to uncover through their geographical distribution and mode of behaviour – and their rhythm of life which follows the polar course of the sun, lead us back in the history of the earth to those regions where, at the beginnings of terrestrial and human evolution, we can look for the Hyperboreans. At that time, everything which was later to unfold as the human race and the kingdoms of nature still lived in a germinal state. Rudolf Steiner described it once in the following way:

> It was a collective world-womb in which the light-plant man lived at that time, feeling himself one with the light-mantle of the earth. In this refined vaporous plant-from, man hung as though on the umbilical cord of the earth-mother and he was cherished and nourished by the whole mother earth. As in a cruder sense the child of today is cherished and nourished in the maternal body, so the human germ was cherished and nourished at that time.[7]

During this period the sun separated from the earth. 'As a result of this departure of the sun, the mist cooled to water.' In this way the 'water-sphere' arose, and the human being was so fashioned that he 'protruded into the mist-sheath, so that he was half a water, half a vapour-being.' There the light reached him from outside, from the surrounding sun.

Now the mystery of the origin of the seals begins to be deciphered. In that region of the earth which is still the home of many seals today, there once lay the cradle of humankind. There the human bodies lived, light flowing all around them, air shining through them, 'on the umbilical cord of the earth-mother'. They condensed and solidified more and more as time passed. Rudolf Steiner, in the lecture quoted above, indicated that at the time when the sun left the earth, man's body had reached that stage which 'we see preserved today in a degenerated form in fish. The fish that we see in the water today are relics of those men.'

But seals are not fish, and yet they are formed and structured for life in the water. If we ask what separates seal and fish from one another, the answer is not hard to find. Seals are closer to man; they are mammals; they form social ties, at least while they are living on land; and they give birth to a single offspring and raise it, if only for a few weeks.

Fish, by contrast, entrust their eggs, often in vast numbers, to the element of water; and even if a few species do build nests and tend to their young for a short time, an infinite distance separates them from the mammals, and especially from the order of seals, in structure, nature and behaviour.

The fish is a water-creature, but the seal-organism is only adapted to life in the water. The limbs are transformed into swimming appendages; the skin is padded underneath with a thick layer of fat; this provides a necessary warmth-shield, and so the round, spindle-shaped body is well suited for swimming. The ear and nose openings can be completely closed in the water. Thus all these characteristics are those of sea-dwellers.

But where do we place the origin of the seals? Were they really once land-dwelling mammals that later went into the water? If this were so, we ought to find at least some indications of earlier stages of these sea-dwellers – but in fact seals appear fully formed in the two last periods of the Tertiary – the Miocene and the Pliocene. According to Wachsmuth's investigations, these geological epochs correspond to the beginnings of the Atlantis era; this means that when the primal forms of the mammals were first beginning to develop, the seals appear already fully formed.[8] Is there not a contradiction here which calls for an explanation?

Might the seals be the ancestors of all the mammals which arose at that time? Not ancestors in the sense of a theory of evolution in which one animal is said to arise from another by those illusory forces of heredity and adaptation, but forerunners in the sense that they have kept their primitive physical form, their rudimentary limbs, and their round bodies, without specialising them? Probably the seals were never really land animals, since it was not until the middle of the Atlantean era that the earth became solid enough for animals and humans to stand and take firm foothold on it.

If we follow up these considerations, we can approach the seal's body anew: is it not reminiscent of an embryonic form? A human embryo at the end of the second month, though not much longer than 25 mm (1 in), has a form and structure very like that of a seal. In the embryo the limbs are still no more than insignificant stumps; the eyes are round, their lids held wide apart. The mouth has no lips – it is like a slit. And the embryo floats in the water of the amniotic sac enveloping it.

Is this something of the early history of the earth? The seals did not become fish, because they stayed within the human family even into the beginning of Atlantean times. They had undifferentiated, embryo-like bodies that moved half floating, half swimming in the still uncondensed water-earth. At the beginning of Atlantis, when memory and language were forming and man's ancestors – among them the seals – were taking the first steps towards development of awareness of the self, the decline of the seals began.[9] They entered too quickly into densification and hardened their embryonic human from.[10] This is why even today they lose their milk teeth at the time of birth and are suckled for only a few weeks. At this time they grow so fast that they very soon become independent. Their hurried childhood is a clear indication of the precipitous process by which they became animals.

The seals bear witness that humankind's first origins lay in those Hyperborean regions that broadly surround the North Pole in the early days of the earth. Here were the ancestors of human beings, and also the ancestors of seals; the two were identical. At the beginning of Atlantis, when the development of consciousness of the personality began to emerge, part of the gradually developing human race, while still in an embryonic form, fell prematurely into solidification. They became the order of the seals. They are the proto-mammals, which became capable of reproduction as embryonic forms. (They are typical representatives of the biological phenomenon called neoteny.)

However, they have kept not only the embryonic gesture of their form but also the inner connection to the sun which once permeated the Hyperborean region. In the rhythm of polar night and polar day, they still follow the sun's course. They were never actually land animals – quite the contrary. Out of the waters of early Atlantis, into which they plunged all too soon, they attempted to take foot on the earth, which was gradually becoming denser. They did not quite succeed. Every year they make this attempt afresh, and in a touching gesture entrust their young to the dry land; but it is only a dream, and passes away as quickly as it came. When the storms of autumn come, they have to go back into the sea, for the setting polar sun is calling them.

That is the earthly destiny of all the seals: as human embryos they densified too soon, and had to submerge themselves beneath the waters of the great oceans. They reached the Antarctic, where they found conditions like those of their former home. Again and again they try to attain the land, and always the water overcomes them. They represent a worldwide memorial to an early stage of human evolution. When we look into their eyes, we see ourselves as we once were, and we sense dimly how we have evolved and what they, the seals, still are.

They are so near to us because they did not become specialised like other mammals. They are neither normal carnivores nor ungulates. They are not forms reminding us of Lemurian times, like the marsupials and the monotremes, nor those forms given to destructiveness which we know as the rodents.

They are related to whales and dolphins, but these have another origin again. Seals later went through the Atlantean catastrophe and thus came into all those regions of the earth which they still populate today.

Mythology

The Inuit once lived in close connection with seals and whales. Formerly, when a seal or walrus was taken by the hunt, the successful hunter stayed in his hut for three days. During this time he was not allowed to take food or drink, nor to touch his wife. All work in his house was left aside; the bedding was not straightened and the blubber-oil not wiped off the lamps.

After three days the soul of the slain seal was free and went back into its mother's womb. Then daily life and hunting could begin again. The souls of the seals, walruses, and whales go home to Sedna, the great goddess:

> [Dr Franz Boas] tells us, the mother of the sea-mammals, may be considered to be the chief deity of the central Esquimaux. She is supposed to bear supreme sway over the destinies of mankind ... Her home is in the lower world; where she dwells in a house built of stones and whale-ribs. 'The souls of seals, ground seals, and whales are believed to proceed from her home. After one of these animals has been killed, its soul stays with the

body for three days. Then it goes back to Sedna's abode, to be sent forth again by her.'[11]

An extraordinary number of rites and taboos among the Inuit are connected with the seals. The souls of these animals are 'endowed with much greater gifts than those of men'. What they cannot stand is the vapour that rises from human blood, and the shadow and dark colour of death. Menstruating women are also intolerable to them.

The Inuit sense dimly that blood and death contain those I-forces and personality motifs from which the seals once withdrew. They still belong to the goddess of the underworld and the earth's depths; they belong to the realm of the mothers, where man's destiny began. Rudolf Steiner said that the 'mothers' are to be found in the bygone stages of earth's development.[12] The souls of the seals reach back to those regions of the past; there they have their common origin with human beings, their brothers.

The latter, however, have denied this brotherhood in recent centuries. In the north and the south of the earth, they began a truly merciless hunt against their own ancestors. Millions of seals, walruses, sea lions, leopard seals, and fur seals have been annihilated. With clubs, axes, guns, and knives whole colonies have been exterminated and species brought to extinction. This was not done by the Inuit, who lived in close connection with the seals. It was done out of greed for money and plunder by Europeans, Japanese, and Americans. Now the seals are becoming even scarcer, and the house of the goddess Sedna must be crowded with their souls. This mother-goddess of the earth's depths also keeps watch over the fate of human beings; but what can become of a humankind that destroys the image of its infancy? Is it not denying its divine origin?

Only a very few people still know the true story of the origin of the seals. Even the Lapp Aslak, son of Siri Matti, who told of the seal's origin on May 24, 1944, 'as the fathers have handed

down,' retained only an intuitive sense of the truth. He spoke
of the exodus of the Jews out of Egypt, of their crossing the Red
Sea, and of the Egyptians pursuing them: 'Then Moses raised
up his staff again, and Pharaoh and all his men, dogs, carts and
horses were swallowed by the waters. Pharaoh himself and all
who belonged to his kind turned into seals; they became big
seals, while all his soldiers became little seals.'[13]

Here again we see, though in a historically distorted form,
an archetypal picture of the seal's origin. One part of humanity,
that does not take part in the progress of evolution but wants to
prevent it, is swallowed by the waters. The other part goes with
dry feet through the sea, and reaches land on the other side.

And at the end of his story Aslak added this comment: 'If you
pay attention when you take the skin off a seal, it looks almost
like a man. Especially when you lay a seal on its belly.' We can
sense how Aslak divined the deep kinship of seal and man.

2. The Life of Penguins

Habitat around Antarctica

In his well-known book, *The Island of Penguins*, Cherry Kearton writes:

> Indeed, I doubt whether, in more than forty years of nature study throughout the world, I have ever found a creature so interesting and withal so amusing. Thanks to the comical expressions for which he is famous, you cannot help laughing at him. But the Island of Penguins has taught me that he is not only to be laughed at. He may not always seem especially intelligent – though he does, presumably by instinct, many exceedingly wise and careful things – but he is virtuous and faithful; and as a model in married life he is supreme.[1]

Whoever has seen penguins – particularly the gentoo penguin – will confirm this description. These birds are strange folk. They cannot fly, but waddle in an upright position on land, or slide forwards on their bellies. In their thousands they emerge suddenly from the sea, to settle at their breeding-places. Then they are like the sand on the seashore. A few weeks after pairing, the female lays one egg. The partners share the brooding, and with touching care they together tend to the upbringing of their (usually single) offspring. All this takes place *en masse*, nose to tail and nest to nest.

After the young have become independent to some extent, the penguins all go back into the sea.

Every year this rhythm is repeated, the only difference being that the various species of penguins have individual rhythms. Most of them breed during the summer. Emperor penguins, however, come ashore in autumn and spend their hatching and rearing time there in the winter. Gentoo penguins even nest twice a year, in autumn as well as in the springtime. However, they also live nearer the temperate zone than most of their relatives.

Penguins are creatures of the Antarctic. Their habitat reaches from the Antarctic continent to the islands and island groups of the surrounding ocean. This entire region is used by the penguins as breeding grounds. It extends from New Zealand across Tasmania to the Kerguelen and Crozet Islands, and South Africa, all the way to South Georgia, South Orkney and South Shetlands. Cold ocean currents even brought them to the Galapagos Islands on the Equator.[2] In the north polar region penguins are unknown.

It is a noteworthy phenomenon that we encounter here, for very few bird or animal species have such a clearly defined dwelling zone. How is this to be understood? Is not the North Pole similar to the rest of the earth, regions where the daily rhythm has become a yearly rhythm because the sun rises and sets only once in the course of twelve months. The region of eternal ice covers both polar zones; but within this cover the two realms are completely different.

> The North Polar region is a great caved-in basin,
> a trough out of which remnants of islands project,
> surrounded by the coastal faces of the continents. By
> contrast, the South Polar region is a continent, an
> extended upland block surrounded by sea.[3]

This characterises the basic contrast between the two polar regions. The following description brings this out even more clearly:

The North Pole is situated in a deep ocean the size of the continent of Europe, and over 4,200 metres [14,000 ft] deep at the Pole itself. The South Pole, by contrast, is the midpoint (at least approximately) of a great continent half again the size of Europe. It lies on a plateau, about 2,900 metres [10,000 ft] above sea-level. Thus, while the enormous land masses of Asia, Europe, and America form an almost closed ring around the deep, wide ocean basin at the North Pole, the continent of the South Pole – the earth's sixth continent – lies totally isolated in a great ocean 5,000 metres [16,000 ft] deep. The Atlantic, the Pacific, and the Indian Ocean flow together into an immeasurable water-mass.[4]

If we try to bring our picture of the two poles into sharper focus, and picture to ourselves that the Arctic Ocean goes down 4,000 metres (13,000 ft) and the Antarctic continent is almost 3,000 metres (10,000 ft) high, so that there is a difference of 7,000 metres (23,000 ft) in height – and if we consider further that in the north we have a sea surrounded with shores, but in the south a giant island surrounded by the sea, then the contrast becomes still more evident in its archetypal clarity. The North Pole is an ocean hemmed in by the shores of the great continents; the South Pole an island swept by mighty oceans.

Sea and island are the two primal forms which mould the landscape of the earth in almost endless variety. The archetypal island is Antarctica. The archetypal sea is the Arctic; and from these two realms comes the formative power flowing into all other regions of the earth. We could almost say that all islands, wherever they may lie, are children of the Antarctic; and the oceans, however small or large, are creations of the North Polar basin.

An island – every island – is a piece of earth which has condensed, crystallised out of the surrounding water. The forces active in the fluid element concentrate themselves at a midpoint and give rise to the island. An ocean, by contrast, is a dissolving process which forms in the centre of the realm of hard earth. From its midpoint the dissolving, liquefying forces flow to the surrounding shores, carrying away cliffs and mountains in the course of thousands of years.

The island is a process of solidification into earth; the sea is a dissolving, a passing away of earth. Condensation and dissolution are the forces active in these two forms. The North Pole is old; there the earth is dissolving. From there the ice ages breathe rhythmically into the course of evolution, covering parts of the northern continents with ice for hundreds of years and then withdrawing to the mother of all seas. From the South Pole, on the other hand, the solidifying island-forces flow into the earth. They hold the continents together and give the earth-realm its hardening properties.

Dissolution flows out of the north, but it is held in balance by the great continental land masses, so that the earth does not become completely fluid. Solidification works from the south; the oceanic water masses oppose this gigantic power, so that the earth does not harden completely.

These are the contrasts between the two polar regions. The wave-like beauty of the seas, which imbues any landscape with a dreamy, intuitive element, since the heavens are reflected in their waters and see themselves there – this comes from the north. The hardening rigour inherent in all islands, which put a stop to the waters and offer to the sky not a mirror but a fist – this comes from the south. Here the earth can stand on its own, and resists the sky.

The penguins are creatures of these forces of self-will. They gather together in places where earth-condensation and island-foundation have their sphere of action.

Fish or bird?

Why do we have a certain feeling of superiority whenever we encounter penguins, and we cannot help expressing this with a slight smile? And when we think of these small creatures carrying themselves upright, why do we feel a sort of tragic-comical sympathy for them?

Is it the caricature of a bird that makes us laugh at penguins? They are birds and also not birds; they cannot fly, and their wings are atrophied stumps that they moves up and down like deformed arms covered with scale-like feathers. When a penguin stretches out his arms, it seems a pitiful gesture; we can see that these stunted limbs could never raise the round body into the air. But the fact that they cannot fly makes these birds not ridiculous but tragic figures.

The comical element has another origin; does it not come from penguins seeming to imitate human beings? Instead of flying, they set himself upright when they walk, begin to chatter and screech, and take themselves so seriously – as if they were really somebody. This is how it seems to us, and so we smile. It is as if a fish, whose fins had turned into feet, were to climb on to land and begin to strut about.

Penguins are actually birds metamorphosed into fish. Their realm is the water, that is where they feel completely at home. In Brehm we read:

> Generally they swim underwater for a distance of about 30 metres [100 ft], then they jump, like little dolphins, up to 30 cm [12 in] above the surface – presumably to breathe – and quickly disappear into the water again. In this locomotion they make use only of their wings; they fly in the water, as it were ... And they move through the water with extraordinary speed – so fast, according to Chun, that they can overtake a steamship with playful ease.[5]

In Gerlach we read:

> With their wings they are no longer able to fly, but they
> make a flying motion with them under water. Their fin-
> wings turn in a rapid, wide swing, up to two hundred
> strokes a minute. The penguins speed along in their
> underwater flight, putting ten metres behind them in a
> second [22 mph].[6]

Thus they can easily swim a kilometre in two minutes, and
thirty in an hour. Is it then so strange that no one can fathom
where they go when they leave the islands with their grown offspring and disappear into the sea? Maybe all southern seas up to
the Equator are their habitat.

When they come on to the land to lay and hatch their eggs
– that is when we first begin to find them comical. As if in
remembrance of their past, they take on the life of birds; the
young males and females find one another, build nests, and a
respectable family life begins. The swift hunter has turned into
a good citizen. The upright carriage, and the striking design and
colouring of coat of feathers – with a white shirt-front covering
his whole abdomen, and a black back like coat-tails – bring out
this respectability even more closely. They stand together in
thousands, chattering, gossiping, pushing one another, taking
each other's stones for nest-building, on occasion even stealing
a wife and her well-guarded egg, but in spite of all they are good
wives and husbands, and faithful parents.

These characteristics, which have been carefully observed and
thoroughly described, make the penguin a comical figure while
on land. It must be a bird, and yet it cannot be one; it lacks wings,
and therefore is bound to the earth. To overcome this lack, it tries
to be like a human being, but this bold attempt fails miserably.
So the penguin lives an unsuccessful existence, condemned to
leave its home in the sea for half the year, to climb onto the land
and live as an in-between form, to recall and repeat its former life

as a bird, while at the same time it stands upright like a human, without being able to be one. Who is not reminded of curses and magic spells enchanting people into the bodies of animals, or condemning human beings to spend part of their existence in places which will be torture for them? We think of how Demeter's daughter is sent down to the kingdom of the underworld, and allowed to come to the light for only a few months. A similar secret lies hidden in the like of the penguins, covered over with a mask of incompleteness and comicality. Has Circe, the powerful sorceress, daughter of Helios, had a hand in this?

Non-flying birds

On the shores and islands of the Arctic there are no penguins. Yet up to the beginning of the nineteenth century birds of a special kind lived there, comparable to penguins: the great auks. Although they belong to a quite different branch of the species of birds, they were subject to a similar transformation. With the great auk, too the wings became stunted limbs and the power of flight was lost. In size they were like a larger species of penguins, measuring about 80 cm (2 ft 6 in). Like penguins, they had a white breast and a black back. Their nearest relatives are the razor-bills and the guillemots of the Scandinavian cliffs.

In earlier times the great auk inhabited not only the islands of the northern seas; prehistoric remains have been found also on the coasts of Denmark and Ireland, and even in southern parts of North America.* The last living specimens were sighted in and around Greenland as late as about 1820. Since then these birds, which may have been as numerous as penguins are today, have

---

* The penguin's name also has its origin in the north. The seamen called the great auk, which they knew, pen-gwyn, white-head. This comes from two Celtic words (pen: head; and gwyn: white), which were later transferred to the penguin, so similar to the great auk. But only in English and German; the French say pingouin for the auk and manchot for the penguin.

disappeared from the earth.[7] In historical times they could be found most often in Newfoundland, Greenland and Iceland.

> All observers mention that they [the great auks] used to swim with heads raised high, but with necks withdrawn, and they always dived when startled. They would sit up straight on cliffs, more upright than the guillemots and razor-billed auks. They walked or ran along with short little steps, upright like a man; and when in danger they dived four or five metres [15 ft] down into the sea.[8]

The great auk brooded in the summer and laid a single egg.

Thus in the Arctic, forms of life similar to those we still find in the penguins had come into being. Both lost the status of birds in giving up the power of flight and in return acquired the capacity to swim. The great auk and the penguin, though probably not related, were subject to the same destiny. Have any other birds had a similar fate? We know of a number of such birds, some still living, others extinct. First of all comes the family of the ostriches. Their wings are atrophied and have such soft feathers that they are useless for flying. The neck and legs, however, are strongly developed, and a fully grown ostrich often reaches a height of over two metres (7 ft). The Australian emu, the South American nandu, and the moa of New Zealand (now extinct for centuries) have a similar build. The moa reached a height of 3.5 to 4 metres (11–13 ft) and had powerful thighs and neck. And we still find, although isolated and rare, the flightless New Zealand kiwi.

All these birds have lost the power of flight. In return they have developed (except for the kiwi) powerful legs and elongated necks. It is as though in these parts of their bodies they had made good their loss of wings. Goethe's poem *The Metamorphosis of Animals* alludes to this:

> Yet within a spirit seems to struggle mightily,

> How it can break the circle and foist its wanton will upon
>   the forms;
> But whatever it may try, it tries in vain;
> For though it penetrates to this limb or to that,
> And furnishes them grandly, still other parts soon wither
>   in return,
> The burden of imbalance destroying
> All beauty of form and all pure movement.

Ostriches and emus are said to move quite gracefully while still young. The older and larger they become, the more the 'burden of imbalance' of neck and legs comes to the fore, and the grosser and more awkward are their movements. The wings and the power to fly were consumed by the legs and the running. In order to keep a balance with the formation of powerful lower limbs, the neck was pushed up away from the body, producing the present forms.

These flightless birds live on the wide-open steppe. Sand, sun, short grass and dryness are their world. The exceptions are the kiwi and the cassowaries of the rain forests of New Guinea and north-eastern Australia. The great Madagascar ostriches (Aepyornithes) were, in Portmann's opinion gradually destroyed because:

> The progressive clearing of the light, savanna-like
> woods of many parts of Madagascar forced the animal
> life of these woods more and more into the swampier,
> less accessible areas of virgin forest ... In these swampy
> wooded regions, the giant ostrich fell prey to crocodiles.[9]

Moreover, the habitat of these birds was the steppe, not swampy woods; for this reason also they died out.

In all these birds the forces of the dry earth predominated. They dried out their bodies, refined their coat of feathers, and the flier became runner. (With the kiwi, the power of the wings went into the elongated, curved bill.) It is characteristic that these birds live mainly in the southern half of the earth.

Are they not real opposites to the penguins? Both groups had their wings cropped by fate. In the penguins, however, the neck and legs were not extended, but drawn in. With every penguin the head sits directly on the breast; the feet stick out like short, absurd stumps from under the belly. Here the air has not dried out the body – quite the contrary: damp and darkness have filled out the body so that the legs and neck disappeared into it. This form reminds us of the whale, the seal and the dolphin. The limbs become little finlike appendages because the bodies, rounded with fat, swell up like balloons.

In the far north as well as in the far south, where cold, darkness and water become overwhelming forces, the great auk and penguin arise as counterparts to the ostrich and cassowary. Are we today still able to recognise the hidden riddle in the existence of these creatures?

The annual cycle

In recent decades the behaviour of large penguin colonies has been thoroughly studied. It has been found that the rhythm of taking to land and departing is different for the separate species. The gentoo penguins studied by Kearton, which live on Dassen Island, north-west of Cape Town, come back on land twice a year, in March and in September; both times eggs are laid and hatched, and the young crawl out. These penguins also tend to lay two eggs while the other species usually produce a single egg.

The emperor penguins, studied by a group of French researchers, choose the Antarctic winter, the bleakest season there is, for brooding and rearing their young.[10] Here it is no longer very convincing to speak of an instinct serving the preservation of the species. The young males and females begin to appear on the high ice plateaus of the Antarctic in April and May. The time of the deep polar night, the period of the worst hurricanes and the iciest cold, they spend without taking any nourishment, tending with touching devotion their newborn young.

As far as we know, most of the other species brood during the Antarctic spring and summer. Thus no uniform rhythm of life can be shown for the penguin family. The different species have their individual periods of migration.

Nevertheless, we cannot doubt that the appearance of the penguins at their nesting grounds is comparable to the return of all other migratory birds. For the penguins, brooding is the time of work and waking. They must suffer through love, birth, and struggle in these months; not until they return to the watery dreamland of the sea will their joy and play begin. What 'the south' is for the birds of our climes, the sea is for the penguins. They are not birds that just happened to fall into water: they have chosen the water as their paradise. They yearn for the water as other birds long to reach their far-away dreamland.

On land, however, they try to behave like all other birds although they cannot fly. They show birdlike gestures and court-ing customs. They build nests of the greatest variety, are monoga-mous for years together, and are especially attentive to their brooding. These avian behaviour forms are exaggerated in the penguins; often they can walk upright, we see kissing scenes in the courting and mating season when the couple will rub beaks and even snuggle their heads together. The young male often spreads out his wing-stumps and tries to embrace his bride; this to the accompaniment of all sorts of noises, screams and barking sounds.

Nest-building is carried out communally by both partners. First one partner works, and when he is tired the other comes, until after some days the enclosure is ready (usually a depression in the ground, like a small cave). Their work done, man and wife climb down to the shore together to bathe and take their evening meal. Then they will meet a few acquaintances for a little chat, stroll along the main street, returning finally to the newly-built house.

The egg, laid after three to four weeks, is greeted with appar-ent amazement and joy, and once again it is both parents who tend the brooding, alternating with one another. This is a time of danger and threat. Rapacious gulls are hovering all around for a chance to steal the precious eggs and make a meal of them. Not more than half the chicks crawl out of the shell, and many fall prey to enemies and the inhospitable conditions.

For the young of the emperor penguins there is no nest. The egg is kept and hatched in an abdominal fold, and the baby penguin squats for weeks on its father's feet, enveloped, protected, and warmed by his abdomen. The mother is far away, somewhere out in the sea, eating to gather new forces. The fathers stand by the hundreds and thousands in the icy polar night, pressed close together to hold off the murderous, annihilating storms. They form whole clusters of penguins, giving each other the last measure of warmth they retain. When the mothers return from the sea, they allow the emaciated fathers their turn to go and to find food and regain their strength. Beneath this bird-cluster crouch the almost featherless baby penguins, waiting for the rays of the sun to return after weeks of absence. The life of the penguins is not all play. When they are on land, they carry their bird's destiny with them as a painful, oppressive memory; misery and need are their lot.

They must once have thrown off their bird-existence to acquire a different, perhaps better, life. So they dive into the ocean, but once or even twice a year they have to give up their enchantment and remember painfully their bird-existence. Kearton describes how the young penguins at a few weeks of age are still quite shy of the water, and only the greatest effort and patience by their parents can induce them to make the first attempts at diving and swimming. Kearton adds something more. He says:

> Sometimes, curiously enough, they seem to imagine that their flippers are wings and that with a little practice they could fly ... At any rate, I have often seen young birds deliberately flapping their wings as if they were quite

certain that that was the way to do it ... This attempt is
so frequent that it cannot, I think, be only the easing of
muscles. Perhaps it is all a legacy of former ages ... when
the penguins really did fly.[11]

Very probably they could, and if so a series of events – or one
event – must have brought about their loss of flight at some time.
After penguins were no longer able to fly, the wings shrivelled
and became fins, which enabled them to live in the water.[12] The
Antarctic forces of island-formation and darkness, the forces of
the great watery wastes, swelled up the body, which absorbed
legs and arms into itself. Thus out of a bird arose a fish; but
a fish that has to be a bird every year, although it is hard and
wearisome. The species would not live on if it did not make the
sacrifice of going on land each year.

The two climaxes of a penguin's existence

During this period on land, the penguin becomes a caricature
of a human being. He even succumbs to a regularly recurring
illness: moulting. Moulting is of course a characteristic of the
entire family of birds, but on most water birds it has the effect
of making them quite unsure in flying for several weeks. Swans,
ducks and geese cast off all their feathers at the same time. They
hide in a thicket on the banks or in a marsh until their body-
covering grows back and they can fly again.

Most birds moult in a less drastic way. The old feathers are lost
little by little and are gradually replaced by the new. During the
process the birds may look wretched, but they can fly and feed.

Penguins, however, are really ill with moulting. For several
weeks they are unable to catch any food for, while some other
birds lose their power of flight during a moult, penguins lose the
power to swim. Kearton reports that the penguins he observed
moult regularly in December. They sense the approach of this

illness for during the preceding days they eat much more than usual in order to have a little supply. 'These penguins are like turkeys before Christmas ... They just go on getting fatter and fatter – and then one day the first definite sign of moulting appears. From that moment all is misery for at least six weeks.' The strength for diving and swimming is lost. The search for food ceases; the bird leaves its nesting place and camps somewhere out in the open. The feathers fall out in patches all over the body. 'You will see thousands of moulting penguins closely gathered on to one of these open spaces, each apparently more depressed than his neighbour.' Soon the whole island is covered a foot deep with feathers, these starving pictures of misery standing among them.

By the end of the moulting the penguins are so emaciated that they are too light to dive under the water. So now they begin to swallow small stones from the beach in order to reach the necessary weight.

> You can see them wandering on the seashore examining pebbles, rejecting some (as being too large, perhaps, or because their sides are not sufficiently smooth for easy swallowing), and gobbling up others till they think that the 'ballast' they have taken aboard is sufficient.[13]

Now the penguin begins to turn into a fish once more. He swallows the stones, hardens and weighs himself down, and becomes a water creature. The moulting period, during which he had to fast and suffer, had given him back to the race of birds. Now the hard stones draw him into the water.

What are the points of climax, the decisive stages, of the penguin's life? Two special experiences are allotted to every penguin. One must be connected with moulting. The condition of distress, illness and fasting remind him of those dark times in which he gave up his existence as a bird and became a fish. For this a yearly penance has been laid upon him – one that all water birds must bear. Is it not remarkable that the penguin does not retreat to spend this period in the dark of his nesting cave? He seeks the companionship of his fellows, as if he sensed that a sorrow shared is a sorrow halved.

Soon after this, however, he eats stones, recapitulating the 'penguin sin'; he becomes a fish, begins to dream, and draws far out into the waters of the ocean.

The other climax is the sight of the newly laid egg. All observers tell with what joy and wonder both parents greet the egg; how they roll it this way and that, look at it from all sides, and simply cannot get enough of it. Is this a sort of intimation of what Rudolf Steiner once said about the form of an egg?

> The egg is nothing other than a real image of the cosmos
> ... philosophers should not speculate about the three
> dimensions of space, for if one only knows rightly where
> to look, one finds everywhere the riddles of the world
> vividly represented. That one axis of the world is longer
> than the other two – a vivid proof of this is the chicken's
> egg; its limits – eggshells – are a true picture of our
> space.[14]

Perhaps an intuitive sense of this knowledge awakens in the penguin parents; they grow to a dawning of this perception. It is this which kindles in them the strength to hatch their egg with the greatest affection and instinctive care, since it has given them this deep experience.

Although penguins on land look so pitiful and ridiculous, the greatness and tragedy of all existence hides under this fool's guise. Penguins are ancestors of the human being; only with them everything is distorted and out of place. For every tragedy has its satire; and so the evolution of humankind has its clownishness in the being of the penguin.

The purpose of a penguin's existence

When Brehm tells us that 'we should not make too much about the benefit' that penguins bring to man, we have to realise that this is a pitiful and spiritless approach. The real question is rather: what is the true meaning of their existence?

We must immediately think of the fact that this group of birds inhabits a continent which has remained untouched by man. The Inuit still live on the edges of the Arctic, but the Antarctic surrounded by ocean, has never been a human dwelling-place. The living conditions are so inhospitable and hostile that men can exist there only for short periods at great sacrifice.

The penguin, however, has the courage to penetrate the dark loneliness and cold of the Antarctic. This alone is a deed of heroism repeated every year. Particularly the adelie and emperor penguins settle on the high snow and ice-covered plains of the South Pole. Are they not carrying a piece of earthly destiny to this abandoned realm of our planet? Then, when the penguins go back into the sea, having finished their brooding work, they swim out in all directions: to the Falkland Islands near South America, to the Cape of Good Hope, to New Zealand, to Tasmania, thus making a living connection with areas inhabited by human beings.

Penguins link the Antarctic continent with the other regions of the earth by acting as yearly messengers between them. It is not national pacts or territorial demands that they disseminate; they bear tidings of the existence of man on earth to the south polar regions.

Perhaps one day a more exact knowledge of the few existing kinds of penguins will make it possible to assign them to the different races of humankind, so that they would represent a shadow image of all peoples existing lands surrounding the Antarctic continent. Even now the penguins, who mimic man in many caricature-like traits, carry his image to the far frontiers of the earth.

Once the whole world of birds was a mighty spirit that moved, brooding, over the waters. For this reason most kinds of birds can still spread out their wings and fly heavenwards. Others had to renounce flight and commit themselves to earth and water.

The penguins are the apes of the bird family. They once had wings; but they fell too quickly into densification, and lost the art of flying. Instead of it they acquired swimming; and now, twice a year they move from sea to land and back again, proclaiming the song of man to the gloomy South Pole.

It is a crude song, more like the braying of an ass than the song of birds. Penguins quack and purr and bark – and yet their voice sounds from within, calling out into the terrible silence of the polar night: There is life on earth.

3. The Migration of Salmon and Eels

Migration

Around the 1940s there was great excitement among scientists, particularly among zoologists and paleontologists, because of a sensational find. Several living specimens of one of the oldest fish in the earth's history, which scientists had assumed to be extinct for about seventy million years, were brought up from the depths of the Indian Ocean. The first one was found on December 22, 1938, but at that time, when the first troubles of the coming war were stirring, not much attention was paid to it. By now [1956] another eight specimens have been fished up. The last one, on November 12, 1954, could even be brought ashore alive and died there since it was not sufficiently protected from the light of the sun.

The name of this ancient group of fish is *Coelacanthidae.* In spite of the 'struggle for survival' and 'natural selection', they have retained the same form as that of their brothers and sisters found as fossils, from the carboniferous period to the cretaceous age, in Greenland, South Africa, Madagascar and Australia. This discovery not only dealt a perceptible blow to Darwin's theory on the origin of species. What is more important is that living witnesses of a very early period of the earth's evolution reach into the immediate present, with no change in form and mode of living, and thus form a bridge which until now had been thought to have perished.

These eight creatures from the primeval world have come up

from the deep seas around Madagascar and South Africa. The next decades will perhaps unveil other such secrets and thus show up how threadbare many of the views are which an agnostic science has formed about the development of organisms.[1] *Latimeria* – this was the name given to the primeval fish – is only one of the signs, which will be followed by many but has also been preceded by a few. Among them is the deciphering of the mystery of the eels.

Only comparatively recently, during the first two decades of the twentieth century, has significant light been thrown upon the migration of eels. As early as the 1890s the Italian naturalist Grassi (1854–1925) identified the larval forms of the European eel and showed that the small fish named *Leptocephalus* which until then had been regarded as a separate species are none other than the larvae from which eels develop. At the beginning of the twentieth century the Danish ichthyologist, Johannes Schmidt, took up these leads and in years of most painstaking research found that the Sargasso Sea in the Atlantic Ocean is the common place of origin for both the European and the American eels. It is their cradle and probably also their grave.

Thus a biological phenomenon could be uncovered which until then had seemed scarcely credible. Eels migrate in regular procession, as they go through their larval stages, from the Sargasso Sea right across the Atlantic until they reach the coastal regions of Europe. This move takes about two to three years. Then they go up the rivers, where they grow big and strong, and after a stay of three to four years or more return again to the sea off the West Indies.

> To the Sargasso Sea!
> Where it is darkest,
> Deepest and darkest,
> There is the goal,
> The beginning and end for us,
> Love and death.

That is how the Dutch poet Albert Verwey (1865–1937) makes the eels speak. And so it probably is, that 'beginning and end' await them there. But why, after years of wandering through the open seas, do millions upon millions of them go up the rivers? Everything which has been thought about this so far has been all too anthropocentric. For instance that the eels found better living conditions in the rivers, or that they were driven to the place where their 'forebears' had been, and other nonsensical notions. Living conditions in the rivers are much more dangerous and difficult than in the open sea; and what does the word 'dangerous' mean to an animal? What does a thought such as 'better or worse living conditions' mean to an animal? What does it mean to an animal to speak of instinct?

The animal lives embedded in a world of facts and experiences of which it is itself a part, which it does not use to greater or lesser advantage, but within which it fulfils the acts assigned to it by a higher wisdom. An animal is never something which is becoming; the human being alone is progressing. An animal is always something which is complete in itself and to which a place is given in a definite environment. Within this fixed existence the self-same act ascribed to this group, family, or species occurs time after time for thousands of years.

The animal acts the role which has been assigned to it for a certain period on the stage of life. Its audience are the gods themselves, who have created this world-theatre out of their own power. Man, too, acts on this stage; but gods watch him, too, and follow his part and sometimes even intervene. But man also beholds his own performance on the stage and knows that he is actor and audience at the same time. The animal is actor only.

Thus *Latimeria* has now been forced on to the stage of this world-theatre; thus the eels entered the limelight and are now seen particularly clearly. There where earthly and heavenly forces meet and play together, they once, many thousand years ago, made the migrations of eels part of the earth's existence.

Now these migrations point back to the evolutionary epoch during which they began. It is there that we should look if we are to gain insight into this strange drama which then appeared in the life of the earth.

But there are all sorts of fishes which migrate; for instance the river-lamprey and sturgeon, salmon and trout; and what has just been said about the eel applies to them all. The salmon, however, show a type of migration and a way of life which is almost the polar opposite of what the eels do. For the salmon, which also travel up rivers and then return to the sea, have their cradle in the head-water region of each river or brook. There the eggs and seed are spread, and the young salmon develop in these places; only thereafter, and at different periods of time, do they go to the open sea. Later they travel back to the places of their childhood, where they then produce their own descendants.

The eels therefore go into the sea to produce their progeny, while the salmon go up the rivers to mate. The eels move from the sea to the rivers and back to the sea. The salmon go from the rivers to the sea and back to the rivers. Thus we have a pair of opposite habits, and a study of their contrasting relationships might yield insight into something which until now has remained veiled.

Contrasting life cycles

Science has coined two names when describing this polarity; the salmon and all those fishes which rise from the sea into the rivers in order to propagate there are called anadromous, and the contrasting eels, which have their cradle in the sea, catadromous. This designation alone would not help much, unless at the same time a living image were to be gained of the phenomena involved. Sufficient details are known today to give us a fairly complete picture of the patterns of migration.

If we could spend a whole year at the mouth of one of the

rivers opening into the North Sea or the Baltic, and were able to observe the fish which move in and out, we should perceive an overwhelming abundance of phenomena. But we must suppose that we could have made our observations in the middle of the nineteenth century, before weirs barred the rivers and factories destroyed life with effluents.

During autumn time, beginning in October and right through the winter, the salmon rise from the sea and go into the rivers. These are the large, mature males and females, which move up towards the river-heads. In the spring these winter salmon are replaced by the so-called spring salmon, almost all males. They enter the rivers until summer and thereafter a quiet period sets in. In late summer and early autumn a salmon is seldom seen at the river mouth. But even so we get varying rhythms for almost every river. Thus in early summer the St James's salmon go up the River Elbe – later the larger females follow, and in the autumn the main army of large, heavy salmon ends the procession.

But at the beginning of May, at the mouth of the Rhine, more or less between May 4 and May 18, the young salmon appear, moving in the opposite direction. One to one-and-half-years old, they migrate to the sea for the first time. After they have spent their childhood up in the mountains they now go out into the great world. They still wear their youthful dress with dark cross-bands which is only gradually replaced by the silvery splendour of the scales. With them, many of the older salmon, larger and smaller ones, swim and drift out to the open sea; they are return-ing to the sea after the spawning season, completely exhausted, striving to regain their feeding grounds.

Always, whether they move upstream or downstream, the salmon go singly; they may find themselves in small groups, but as though by accident. With the eels, it is quite different. Coming from the Atlantic, they appear on the west coast of Europe in late winter and during spring – earlier in Ireland and England, and correspondingly later in Denmark, Germany and the Baltic countries. Tens of thousands of small, transparent elvers, 6 to 8 cm (about 3 in) long, form long processions which move up the rivers. An observer described it as follows:

> One morning at the end of June or beginning of July, when we went on to the dyke of the village Dreenhausen which looks directly on to the River Elbe, we saw a dark streak which moved along the whole length of the river bank. It was formed by countless numbers of young eels which moved up river near the surface of the water and always kept so close to the bank that they followed all the bends and coves ... This marvellous procession of fish continued without interruption or diminution throughout the day and the following day as well.[2]

The eels which go upstream remain in the rivers and brooks for several years and then, having become silver eels, large, darkly pigmented and round, they return to the sea. Mostly in autumn they move back into the open ocean; again this does not happen singly but in smaller and larger processions.

Almost throughout the year the river mouths become the gates of entry and exit for billions of fishes which keep a living communication between the salt water of the sea and the fresh water of the rivers and streams. With tremendous forces both eels and salmon strive up river. To quote Brehm:

> At the end of June, I was at Ballyshannon in Ireland at the mouth of the river which in the preceding month had been flooded. Near a cascade it was clouded by millions of small eels which constantly strove to climb the wet rocks on the edges of the waterfall and died in their thousands; but their damp slippery bodies made a ladder for the others and enabled them to continue on their way. I even saw them climb vertical cliffs; they twisted through the damp moss or got a hold of the bodies of others that had died in the attempt.[3]

Similar, but not so massed, is the upward striving of the salmon. They overcome the greatest obstacles, such as rocks and waterfalls, by flinging themselves from rock ledge to rock ledge, step by step coming closer to their goal.

Eels and salmon move and swim through the entire length and breadth of a river; they enter most of the tributaries, the streams and streamlets, so that they are intimately interwoven with the whole network of a single river. The eels seem to like broad expanses; the salmon prefer the upper reaches. The river as a biological unit is filled with these fish and through them finds a close connection with the sea.

The life of salmon and eel, however, differs not only in that the one are anadromous and the others catadromous, for this polarity also expresses itself in many individual traits.

Eels stem from the Sargasso Sea, breeding where the Gulf Stream coming from the Gulf of Mexico, turns north off Florida and flows along the coast of North America. It is a huge region, almost the whole of which is washed by an arm of the Gulf Stream. The sea is about 6,000 metres (20,000 ft) deep there, and from their spawning grounds at around 400 metres (1,300 ft) deep the young eels rise up. The American species breeds farther to the west, the European one farther to the east of this region. Billions of tiny, transparent, leaf-shaped fish, 2 to 3 cm (1 in) long, rise from these depths. Their form is still like a real fish form which, as they approach the western and eastern coasts, increases in size. The migration to Europe, which at the same time means the metamorphosis into elvers, takes two-and-a-half to three-and-a-half years. It is assumed that fertilisation takes place between March and April, at a depth of about 400 metres (1,300 ft), and that the young larvae gradually rise upwards, grow and – in the course of their eastward migration–change into the round, wormlike elvers. The westward procession is shorter and the American elvers are therefore smaller than the European.

As soon as they have reached the rivers their bodily functions change. A yellowish brown pigment deposits itself in the skin and at the same time a tremendous intake and elimination of substances begins. Then their growth is rapid and they often gain enormously in size and weight. They stay in the fresh water region for three to eight years. During the day they remain hidden in the mud of the river or stream bed and only after dark does their hunting time begin. The eels are especially active at night; they fear the daylight. Louis Roule, one of the most knowledgeable ichthyologists, writes:

> [The eel] is to be numbered with the nocturnal creatures of the waters which become active only when it is dark. It retains the impression acquired in the depths of the sea where it was born ... If eels are kept in an aquarium and a beam of light is suddenly brought to bear upon them, they get into a panic immediately, dash off in the darkest corner, pressing as close to one another as they can in their eagerness to get away from the light they hate ... This persistent, inescapable dread of the light is a controlling factor in the life of the eel and makes it obvious to us that the eel is indeed a creature of the depths.[4]

But it is not only the light that they fear. They withdraw also from the cold and during winter they are inactive; buried in the mud they go through a kind of hibernation. Among all the river fish they are the first to bury themselves in the autumn and the last to reappear in the spring. Their life-element is dark warmth; cold brightness is the realm from which they flee.

The salmon are quite different. When they rise from the sea into the rivers in autumn their whole body begins to shine. Gerlach describes it thus: 'As they go up, the salmon change colour. Red spots glow on the gill covers and sides of the males and are also strewn over the bluish shimmering head. The belly turns purple. A rosy pink colours the fins.'[5] Clad in these glorious hues they climb up the river until they have reached the springs of the various tributaries and brooks. They aim for the light and cold in the heights; around and after Christmas, in the ice-cold and light-filled water, mating takes place.

The females pour enormous masses of eggs into a hollow they have prepared beforehand with their fins and the male sperm is shed over these. This mating lasts for one or two weeks and soon afterwards most of the exhausted fish die and few make their way to the sea whence they came. A year later they will spawn for the second time but rarely a third.

During the whole journey up the river they have not taken any nourishment. The intestines are as if degenerated at this time and cannot hold food. Instead the reproductive organs develop to a huge degree. By the end of the journey the ovaries are almost a quarter and the testicles almost an eighth of the complete body weight.

The hatchlings (fry) from the fertilised eggs soon develop into parr and stay for one to one and a half years in the vicinity of the source. Only after this time do they make the journey as young salmon – smolt – down the rivers and back to the sea. Then hunting begins in their two to four year life. During this period in the sea they wander through a wide area. Salmon that had been marked off the European coast were found off Kitaa, western Greenland. They seem to prefer coastal zones and hunt mainly in the upper 10 metres (30 ft) of water but do also penetrate the depths.

The eels, however, that are plump and big by the time they get back to the sea are no longer predators but have one singular goal: to get to the Sargasso Sea! And now these creatures who loved the darkness begin to develop larger eyes; they become eight or nine times larger and at the same time their darkened body begins to get a silvery shine. Perhaps they still cannot bear the direct light but reflected sunlight that is broken by the water and that now leads them the way back to the depths whence they have come.

Thus salmon and eels really are polar opposites. Salmon love light and coldness; eels fear them, but love the warm darkness in which the salmon have no part. Both of them link oceans and rivers, salt water and fresh water, and live in the circulatory system of the waters of the earth.

The influence of light

The contrast between salmon and eel is also bound up with their different life-elements: the salt water of the sea and the fresh water of the rivers and streams. And how great must be the difference between those generations which have their fertilisation and early development high among mountains and hills near the springs, and those who experience this beginning of life in the dark depths of the ocean! We have already mentioned that eels spawn at a depth of 400 metres (1,300 ft). If one assumes 1,000 metres (3,300 ft) as an average height for the breeding grounds of the salmon, the difference in height between these two beginnings to life is evident.

And the breeding season is winter for the salmon, in contrast to summer for the eels: furthermore, the regions sought by the salmon are northern hills, whilst the Sargasso Sea off Florida belongs to the sub-tropical region.

The eggs of the eel therefore develop in the darkness of salt water, those of the salmon in the light of fresh water. Here far-reaching qualitative differences play a fundamental role in the development of the fish. For sea water not only contains considerably larger amount of salts compared with fresh water, but the special composition of its chemical elements deeply influences the organic life which is constantly growing in its womb. The fresh water of the rivers and streams is just the opposite; it is brighter, lighter and quicker, and certainly does not have the brooding character which is so typical of sea water.

Rudolf Steiner once gave a fundamental characterisation of this contrast:

Yes, you see, if one really investigates sea-water, one discovers that this salty sea-water stands in but slight connection with the universe ...

The springs with their fresh water are open to the universe, just as our eyes look freely out into space. We can say therefore that in countries where there are springs, the earth looks far out into the universe; the springs are earth's sense organs, whereas in the salt ocean we have more the earth's lower body, its bowels ... And everything through which the earth stands in connection with the cosmos comes from fresh water, everything through which the earth has its intestinal character comes from salt water.[6]

This remark can open up a direct understanding of the phenomena we just described. For it is obvious that the salmon lay their eggs into the 'eyes of the earth,' in the springs of rivers and streams and thus right from the start are intimately bound up with the light of the universe. Hence their display of colours, their brightness, their 'salmon-coloured' rosy flesh, their tremendous strength when they move upstream in the rivers. It is the universal light which permeated them and can now live and act in their bodies. They are fish woven from light, and when they move from the hills into the seas, take this light with them and carry it into the wide expanses and depths of the ocean.

Eels, coming from the low-lying regions of the salt water, remain bound to darkness. They are nocturnal animals which flee from the light and strive towards darkness. Colourless, that is, without light, their larvae rise to the surface of the ocean, drift towards the shore, and lose the characteristic fish form in order to metamorphose into the snake-like form. But as soon as they touch fresh water their surface acquires a yellow pigmentation, so that the light of the new surroundings can be reflected. When the eel grows up, it turns dark yellow underneath and black-green above and spends many years as a water snake in the rivers.

Just as the salmon is exiled to the sea in order to bear its bright light to the depths, so the eel is assigned to the rivers, to bring to them a necessary element of darkness.

But when at the end of its 'river time' the eel returns to the sea and moves homewards, then its eyes begin to grow and the body shows a silvery shine. For now it needs the reflection of light which until then it had shunned. It needs it in order that its sexual organs may acquire the power of reproduction. That which is given to the salmon by the light-filled springs and streams the eel acquires through the enormous enlargement of its eyes. Both need the light so that their progeny may be assured.

Rudolf Steiner speaks also of the salmon and describes how necessary it is for these fishes to go up the rivers, in order to receive the celestial forces for their reproductive organs:

> Gravity is the earth-force and works upon everything muscular, everything bony. The earth shares its salt with us and we get strong bones and muscles. With this salt excretion of the earth, however, we could do nothing for our senses and the reproductive organs; they would wither away. These must always come under the influence of extraterrestrial forces, the forces coming from the heavens. And the salmon shows what a distinction it makes between fresh and salt water. It goes into salt water to take up earth forces and get fat.[7]

Thus we gain a more complete picture of the phenomena we are studying. We understand now the polarity at work in the sea and in the rivers and learn to see that the migrations we have described have a meaningful background. They carry the light of the rivers and streams into the darkness of the seas, and the darkness of the ocean into the brightness of the fresh water regions. Eels and salmon are the constant bearers of a form of breathing which makes light and darkness flow to and fro between sea and rivers. From the grotesque deep-sea forms produced by the darkness to the bright vesture of herrings and sprats, trout and minnows – which act almost like arrows of light – brightness and darkness play into each other.

Salmon and eels are taken hold of by this light and darkness; they absorb it and become its messengers.

If the idea we have now gained is fully visualised, we come closer to understanding why a salmon, when it goes up the river, always goes back to the very place where it grew up. At first this was explained as being due to heredity but more recent experiments and observation prove unequivocally that the salmon does not return to the place where its forebears lived, but to where it spent its youth.

Gerlach reports on breeding experiments in America which prove this most impressively:

In the Columbia River in the north-western United States it has been possible since 1939 to resettle the Chinook salmon. These go up the Columbia River. After they have swum 600 km [400 miles] the Grand Coulee Dam now bars their way. It is 180 m [600 ft] high and the water in the reservoir above is too warm for the salmon. In 1939 the Federal Fish and Wildlife Service, under the direction of Dr Ira Gabrielson, began to catch the salmon 120 km (75 miles) below the Grand Coulee Dam as they came up the fish ladders and to take them in big tank wagons to the breeding station at Leavenworth, where the eggs were artificially fertilised. When the hatched fry were big enough to use their fins they were taken in tank wagons to the upper reaches of the Wenatches, Eutiat, Okanagon and Methow rivers and put out there. These rivers flow into the Columbia river below the Grand Coulee Dam. The young Chinook salmon remained there until after one year they had reached a length of about 15 cm [6 in]. They then drifted down to the ocean and were seen no more. They should have reappeared in 1944 to spawn, five years after they were born. And they did it. When they were small they had been marked by cuts in their fins. Those which had been placed in the Wenatches took without hesitation the way to this river, and the salmon from the Eutiat, Okanagon and Methow were equally sure in finding again the place where they had grown up. The home of their parents – above the Grand Coulee Dam – they did not seek any more. The same procedure was used in later seasons, and the resettlement experiment was successful.

This indicates precisely the strange ability of the salmon to find their way back to the place of their origin, the fountain of their youth, after many years of travel which must take them thousands of miles. To write of such an ability as 'instinct' would be dodging the issue. Heredity cannot play any role in it either; it is a different principle which can be conjectured. The fish cannot 'find' the way; they have neither the sensory nor the discerning faculties for such a complex ability. The only possible consideration is that they move 'blind' but 'blind in the sense that aeroplanes can fly 'blind', because they are on autopilot. It is only when we begin to imagine that every single salmon remains linked to the special light of the place of its youth, and that a delicate beam of this light accompanies it on all its travels, invisible to the human eye but perceptible to the salmon, that one comes closer to a first understanding of this strange phenomenon. Like the two children in the fairy-tale who drop breadcrumbs on their path through the thick forest so that they can find the way home again, the salmon leaves small threads of light behind in the sea, so that even years later it can follow them back to its place of origin.[8]

To me it looks almost as if the millions of salmon which every year descend from the rivers into the darkness of the sea carry these flickers and threads of light with them and thereby gently lighten the darkness of the ocean. When they return to their river they roll this thread of light up again and use it for the colours of their wedding garment. This concept can be borne out if one remembers the many different deep sea fish which are equipped with strange little lanterns and lights on all sorts of parts of their bodies. In their case these sources of light are visible also to the human eye, because these fish do not spawn high up at the springs and thus do not receive the invisible light in the form of light-ether as their endowment at birth. The deep-sea fish have to create these light sources from within their bodies and thus it becomes coloured light which glimmers but is not able to shine.

A similar riddle is provided by the eels, which in their migrations find the rivers and on their return never miss the Sargasso Sea. Here again 'blind' swimming can provide a plausible explanation. The eels come from the dark depths and rise to the surface of the sea as transparent larvae floating towards the European or American shores. In their nursery, however, no distinction is made between American and European larvae: the two grow up together and yet every single larva 'knows' where it belongs. So some float westward, the others eastward.

If one were to speak of instinct one would again only obscure the issue. But it can be visualised that both the European and American rivers radiate a very delicate light into the darkness of the sea and that different light qualities shine from the east and from the west. Almost the only difference between the American eel larvae and the European is that the first have fewer vertebrae: the former have between 103 and 110, while the latter have between 111 and 119.[9] It is perhaps possible that the spine acts as a very delicate antenna which draws one group to the west and the other to the east.

In such problems of animal life one must not overlook the environment and its wide differentiations. Its manifold forces act upon the animal which has been created out of them and for them. An eel larva is almost nothing but a complex sensory organ and only when it changes into the rounded form of the elver does it develop an additional digestive and metabolic organisation. Thus the 'light antenna' – the early larva – is guided in the direction from which the light which it is able to perceive is shining. Near the river this light trail becomes so strong that the skin becomes pigmented and the animal metamorphoses into a feeding organism.

Years later, when the fully matured eel again travels from the river into the sea, the eye begins to grow; for now the eel returns to its dark homeland and again turns into a sense organ. Now it ceases to eat and yearns for the darkness whence it has come. Both salmon and eel spend their lifetime in a continual interplay of light and darkness.

Thus we can understand what Rudolf Steiner once said so vividly about the nature of fishes:

> The fish has water within it, yet the fish does not feel itself as the water; the fish feels itself to be what encloses the water, what surrounds the water. It feels itself to be the glittering sheath or vessel enclosing the water. But the water itself is felt by the fish as an element foreign to it, which passes out and in, and, in doing so, brings the air which the fish needs. Yet air and water are felt by the fish as something foreign. In its physical nature the fish feels the water as something foreign to it. But the fish has also its etheric and astral body. And it is just this which is the remarkable thing about the fish; because it really feels itself to be the vessel, and the water this vessel encloses remains connected with all the rest of the watery elements, the fish experiences the etheric as that in which it actually lives. It does not feel the astral as something belonging to itself.

> Thus the fish has the peculiar characteristic that it is so entirely an etheric creature. It feels itself as the physical vessel for the water. It feels the water within itself as part and parcel with all the waters of the world. Moisture is everywhere, and in this moisture the fish at the same

time experiences the etheric. For earthly life fishes are certainly mute, but if they could speak and could tell you what they feel, then they would say: 'I am a vessel, but the vessel contains the all-pervading elements of water, which is the bearer of the etheric element. It is in the etheric that I am really swimming.' The fish would say: 'Water is only Maya; the reality is the etheric, and it is in this that I really swim.' Thus the fish feels its life as one with the life of the earth. This is the peculiar thing about the fish: it feels its life as the life of the earth, and therefore it takes an intimate part in everything which the earth experiences during the course of the year, experiencing the outgoing of the etheric forces in summer, the drawing-back of the etheric forces in winter. The fish experiences something which breathes in the whole earth. The fish perceives the etheric element as the breathing process of the earth.[10]

Here Steiner describes the extraordinarily subtle sensory world of fish; as sensory antennae they perceive the breathing of the earthly ether-world. Eels and salmon are embedded in this ether realm; they swim in the water; but they live in the etheric flow of light and darkness, of warmth and cold, of sound and chemical activity, of living and dying.

Therefore salmon are connected to the course of the year as a sundial is to the progress of the day. In the autumn, when the earth-ether is inhaled, they go up to their springs, to spawn at the time of midwinter. In the spring they migrate into the sea, in close connection with the exhaling stream of the earth's ether.

Thus too eels return to the sea in autumn, to spawn in the Sargasso Sea at midsummer and then to die.

Both species of fish live in the interplay of light and dark which continually recurs between the sea and the rivers, between salt water and fresh water, between the cosmic forces and the depths of the earth.

The spiritual evolution of salmon and eels

Since Johannes Schmidt made the important discovery that the Sargasso Sea is the breeding place of the European and American eels, many scientists have given much thought to the problem why this happens just there and nowhere else. More and more it was seen in connection with Wegener's theory of continental drift and today a group of scientists assume that the eels were once fresh-water fish which – as it were – fell into the sea as the European and American continents gradually separated.

Muir-Evans for instance, describes how the Afro-European land mass separated from the American block and then writes:

> If this is correct one can visualise the larva of the European eel as originally ascending eastern rivers which were quite handy, but as the continents became separated the journey of the larva to its fresh water would gradually be extended until a journey of three thousand miles [5,000 km] intervened. It is difficult to think geologically but this theory alone gives reasonable explanation of the migration of the European eel.[11]

Eugen Kraus too tried to solve this problem. He was the first one who pointed out that the migrations of the eels seem to be connected with the Gulf Stream. He thinks that Rudolf Steiner's statement that the Gulf Stream once flowed around the Atlantean continent may help in the solution of this important riddle.

However, if Rudolf Steiner's indications are studied accurately, the former Atlantean continent does not in any way come near the region of the Sargasso Sea of today. For the statement reads:

> This continent was encircled by a warm stream which, strange as it may seem, was seen clairvoyantly to flow

from the south through Baffin Bay towards the north
of Greenland, encircling it. Then, turning eastward,
it gradually cooled down. Long before the continents
of Russia and Siberia had emerged, it flowed past the
Ural mountains, changed course, skirted the eastern
Carpathians, debouched into the region now occupied by
the Sahara and finally reached the Atlantic Ocean in the
neighbourhood of the Bay of Biscay ... This stream is the
Gulf Stream which at that time encircled the Atlantean
continent.[12]

This shore-line of the old Atlantis, the centre of which
was roughly in the region where Ireland lies today, leaves the
Sargasso Sea far outside its boundary.

The distribution of the eels, however, runs only partially parallel to this shore-line. Kraus himself says:

The European eel therefore appears on all European
coasts, along Scandinavia up to the White Sea and in the
Baltic Sea, in the Mediterranean and on the north coast
of Africa, in the Black Sea, from the Sea of Marmara to
the Sea of Azov, and finally to Iceland (except on the
north and north-east coast).

The American eel settles in the eastern regions of
North America up to Labrador, in south-west Greenland,
the north coast of the Gulf of Mexico, the eastern edge
of the Caribbean including the Bahamas, the Greater and
Lesser Antilles, down to the Mouths of the Orinoco on
the north-east coast of South America.

This delineates the region in which the eels migrate today,
but it is very different from what was once the outline of Atlantis
in the way described by Rudolf Steiner. It may well be that the
eels did once go up the rivers of Atlantis and that, as that continent gradually subsided, they gradually moved to other coastal

regions. But the Sargasso Sea must have been the uterus of the eels since primeval epochs, long before even Atlantis existed. They are fish belonging to the most ancient days of the earth.

They have, however, retained their peculiar metamorphosis from fish to the form of a snake, similar to the developmental stages of the frog and other amphibians which still show their origin from fishes as tadpoles. This transformation of the larva of the eel into its snake-like form is no doubt a pointer to very ancient bodily changes to which this creature must have been subjected.

Anyone who has studied anthroposophy in relation to earth-evolution can hardly doubt that this transformation of the eels occurred during the period when humanity experienced the biblical Fall. It is the same period during which the moon separated from the earth thus making other steps in the course of evolution possible.

Like all other fish, eels must have their archetypal origin in the Hyperborean age, during a time when sun and earth were still one heavenly body, permeated with light and warmth. But when the sun separated from the earth, the fishes were born as animal beings. Rudolf Steiner indicates this:

> [Man] had only a very fine etheric form at the time when the earth and sun were still united. When these separated he thrust from him certain animal forms, and these have remained behind at the stage in evolution which corresponds to the time when the sun was still within the earth. From these, entirely different forms have naturally arisen in the course of time ... Were we to select a characteristic form which is still to be found today, and which may in some way be compared with those which remained behind when the earth was thrust away by the sun, we must select the form of the fish ... it is that which still has within it the last echo of the sun-forces.[13]

This throws a first but very important light on the evolution of the fishes, showing that they must be counted among the oldest earthly organisms. It is from this primeval time that *Latimeria* has survived unchanged until our day.[14] Other fishes, probably most of them, underwent further evolution and the metamorphosis of the eel from a tiny fish into a kind of snake is but *one* instance of such transformations.

In the same lecture Rudolf Steiner said:

> When the sun separated from the earth, the earth went back in development; it degenerated; and only after the moon withdrew with the worst constituents did improvement again take place. There was, therefore, for some time an ascending development until the departure of the sun; then a descending one, when everything became worse, more grotesque; then, after the moon withdrew, a re-ascending development again.
>
> From this stage of evolution we have also a form which has degenerated, and which does not by any means appear now as it did then, but it exists; it is the form which belonged to the human being before the moon withdrew, before he had an 'I'. The animal form which recalls the lowest stage of earthy development, the time when man plunged most deeply into passions and when his astral body was susceptible to the worst external influence, is that of the serpent.[15]

The eel has retained this form and it is still possible to follow the changes which lead from the form of the fish to the form of the snake, via the elver to the yellow eel and to the shining black eel. The eel is like a living memorial which has been preserved to remind us of that period when the form of the snake developed into the reptiles of today.[16]

Since that time, too, eels have become carriers of poisonous substances. Their blood and body fluids have a paralysing and lethal action on other organisms. This is a remnant of that lowest point in earth evolution of which Steiner speaks.

In the legends and myths of the peoples of Madagascar, Australia, the Philippines and the South Sea Islands, eels play a great role. The souls of the dead live in them and ancestors are even called eels. This is a region inhabited by a large eel family. Kraus writes:

> There are nineteen species, three of which belong to
> the temperate zones and the other sixteen to the tropical
> zone. In the northern temperate zone the Japanese eel
> is to be found, which is strikingly similar to the Atlantic
> species and therefore stands out clearly among the other
> Indo-Pacific species.

Here, where once the moon left the earth and where the Lemurian continent lay, is another eel region perhaps as important as the Atlantic one. It is not yet known where in the Indian or Pacific Oceans the breeding place of these Lemurian eels lies which would correspond to the Sargasso Sea. But that it is to be found in the very region where the metamorphosis from fish into eel occurred is significant and Kraus was the first to indicate this emphatically.

The salmon, however, has remained free from this world of the Fall. It lives in the light of the springs, where it has its youth, and moves from there to the sea as a bearer of light. It is a true fish which points back to that Hyperborean time when sun and earth were still one. A remnant of the brightness which permeated that earth epoch still lives in the salmon today. The area of its distribution is quite different from that of the eel. Brehm writes:

> The home of the salmon must be taken to be the waters
> of the temperate region of Europe, southwards down to
> a latitude of 43°N, and of the New World down to 41°N.

This fish is absent from all the rivers which run into the Mediterranean. In northern Europe it visits chiefly the Rhine and its tributaries, the Oder and the Vistula, though it is not absent from the Elbe and Weser. It is found more frequently in the rivers of Great Britain, Russia, Scandinavia, Iceland, Greenland and North America, more rarely in those of western France and the north of Spain.[17]

This means a belt which envelops almost the entire Arctic region. Like a ribbon spread around the North Pole the salmon live in both the Old and the New Worlds. And thus they have remained faithful to the region in which they first arose. The Hyperborean region was where the salmon are today, in the north, where light and cold act in a direct way upon the earth's surface.

The salmon has remained faithful to the sun. When this great star separated from the earth, the salmon followed it by rising from the sea upstream into the northern rivers, higher and higher towards the light of the sun, so that at the furthermost point, where the springs become the 'eyes of the earth', it could be closest to its creator. There it performs the sacrificial deed of procreation and its young remain for more than one year devoted to the realm of light.

The eel, on the other hand, followed the path of the earth. Eels went down to the depths of the Atlantic and Indian Oceans and thus left the light and became the sons of darkness. Hence they had to transform into snakes and become bearers of the darkness. Eels came under the forces of the moon, and are still ruled by the moon today as they were once upon a time. Their colour is yellow and greenish, similar to that of the waxing moon. Like the invisible new moon they rise from the depths as transparent and insignificant forms, increase in strength and travel up into the rivers. But even there the eel remains a nocturnal animal, in thrall to the moon.

Both salmon and eel are fish. But the one has remained a

pure fish and thus a son of the sun; the other has developed further, become a scion of the moon and takes on the destiny of the snake.[18] Yet they are brothers; they know of each other and one day, when the end of the earth has come, they will return together and become children of God.

At the end of the Sermon on the Mount (Matt.7:9f), Christ says: 'Or what man of you, if his son ask him for bread, will give him a stone? Or if he asks for a fish, will give him a serpent?'

In these words from the New Testament lives the statement which shows fish and snake, salmon and eel, in their mutual relation as brothers. Rudolf Steiner brings all this together when he said:

> Fish and snake symbols are derived from the mysteries
> of our evolution. It is quite natural for a person to
> experience a feeling of pleasure when he sees the
> glistening body of a fish in the pure, chaste water
> element; it gives him a feeling of peace; just as to those
> of a pure disposition it gives a feeling of horror to
> see a creeping snake. Such feelings are by no means
> meaningless memories of things once passed through.[19]

The witnesses of this are the salmon and eels. They are the living witnesses of a past and yet extant world which to our senses is hardly perceptible, but can be revealed by imaginative understanding. It is the sphere from which all living beings arise; that kingdom where the powers of light and darkness weave and flow and are the fountain-head of all creatures.

Year after year the teeming eel larvae rise from the depths of the Sargasso Sea and the Indian Ocean; the ocean transforms its creatures from fishes into snakes and then calls them back again.

Opposite to this welling up from the depths are the thousands of springs of northern rivers, streams and brooks in America, Europe and Asia. There billions of eggs of the salmon – spread out in a huge circle around the globe – are annually hatched by the light; from the periphery the young salmon move to the centre around the Arctic Sea.

Here are two circulations which stand opposite each other: one begins in the centre and the other in the periphery; one bears light, the other darkness. Interweaving, they maintain the circulation of all the waters which are filled with light and darkness.

> And there is no thing in nature
> Which has been created or born
> Which does not reveal its inner form externally.

(Jakob Böhme)

4. Elephants

One may regard knowledge as something real only when one says to oneself: that which I now create in my spirit as knowledge must first be contained in the things. And inasmuch as I take something of the things into my spirit, regardless of which realm they belong to, I presuppose the spirit in all realms.

Rudolf Steiner[1]

The image of the elephant

Elephants appear in our consciousness from our earliest childhood. Their image and presence accompany us as we grow up and begin to learn. First we meet them in the circus and in the zoo. Then as children we gaze up in awe and astonishment at the huge mass which towers above us. We are fascinated by the play of the trunk and are perhaps terrified by the snorts and trumpeting which may suddenly ring out.

Later we read in adventure books about the life of these mighty animals in Africa and Asia; we hear of the extraordinary strength of the working elephants in India and Burma and we see pictures of them in festive processions, decked in wonderful materials, when they used to carry maharajas on their backs.

Later still we learn of the use of elephants in the Punic Wars and of the panic among the Roman legionaries at the Battle of Tunetum, when the Roman forces were destroyed. We travel with Hannibal's elephants as they journey over the Alps, only one of them finally setting foot on the plains of Italy. And then we read how the Romans gradually became familiar with war elephants and began to use them.

Thus the elephant became an animal familiar to everyone. Not exactly a domestic animal such as cows and pigs, dogs, sheep, and goats, but closer than bears and snakes, which are near to us and yet so far. The elephant is tamed only in Asia; in Africa it remains undisturbed and beholden to no-one, at least in so far as it is not a victim of hunters, poachers, and profiteers.[2]

Somewhat as the elephant accompanies us as image and experience through childhood, so it took part in the waking of humankind. It was there when humans began to tread firmly on the earth, and in prehistoric times it was at home in most parts of the world. It inhabited south-east Asia, America, Europe and Africa, from the Arctic Circle to the Equator and further south. It seems only to have missed Australia, New Zealand and South America. As mammoth in the north, as elephant in the south, the whole group of the Proboscidea (mammals with trunks) was spread out over 352 types.

Today this multitude of types has dwindled to a very few. There is the African elephant, which lives in the central territories of that vast land-mass, and has two sub-types: the steppe and the jungle elephant (*Loxodonta africana africana* and *Loxodonta africana cyclotis*). The African elephant is larger and stronger, and has mightier tusks, than his Asian brother, and seems to go further back into the history of the earth. All other species, with one exception, are now known only through imprints and skeletal remains. The exception is the mammoth, which in a few instances has been found in northern Siberia as a fully preserved animal complete with skin, hair and organs, even with food in its stomach – thousands of years after the great beast had been frozen to death. Its blood was thawed out, and serum tests showed that it really was a close relative of the present-day Indian elephant. Thus do prehistory and the present day touch hands.

We find this within ourselves too. We find in the depths of our souls, where the powers of the will are born, a connection with these animals. They belong to us, are part of our existence and substance. As children we gaze up at them; as young adults we would like to meet them and to understand them; as grown-ups we turn away from them, because we are afraid of meeting their existence in our own depths. Only a few groups of people remain close to them and stay connected to them throughout their lives. Once the whole of humanity lived in a kingdom ruled by elephants. These creatures ruled the earth; they possessed lands and territories, and human beings became their master only gradually. The elephants withdrew and lost their kingdoms and estates; now they tread the path of their existence as perpetual beggars, and yet are kings.

Physiology and appearance

'The way in which a being lives outwardly shows what it directly experiences inwardly.'[3] Rudolf Steiner framed this far-reaching statement when he was speaking of a possible understanding of the animal organism. We will now look at the appearance of the elephant.

We will take the ancient African elephant and approach him from the front. He looks at us with small deep-set eyes. The forehead rises like a cliff and spreads out into the upper edge of the great wing-like ears. The ear is approximately one metre long and 80 cm wide (40 by 30 in); it encircles the skull and the throat like a huge umbrella separating the head from the body. Seen from the front, the body is so to speak hidden behind the enormous ears and only the legs, like giant pillars, are in our field of vision.

The face is elongated from the forehead to the great trunk, which reaches to the ground as if it were a fifth limb. This organ is a highly developed upper lip which has become fused with the nose and covers the lower part of the expanded mouth. On either side of the trunk the two tusks rise in a concave curve from the upper jaw.

The trunk is the symbol of the elephant. His nature and being, as it were, are embodied in it. What is this strange apparition?

The elephant tears leaves and twigs with it from trees and bushes, and stuffs them into his mouth. He does not gather food with the mouth itself, as do hoofed animals and carnivores. He tears out huge tufts of grass with his trunk, and then, having knocked the soil away on his knees, he consumes the mouthful. He sucks up water with his trunk and squirts it into his mouth. The trunk is therefore a limb which connects him with the outer world, as do the arms and legs in humans.

The elephant deals hard and soft blows with the trunk. He places his trunk lovingly around the rump of a fellow. A young elephant will wind its trunk round the mother's tail when moving around and is thus guided along.

The trunk, however, is more than a limb. It is at the same time an extraordinarily fine and perhaps all-embracing sense-organ. A large proportion of the elephant's sense capacities are concentrated there. The elephant has a wonderfully keen sense of smell; some observers have suggested that a good supply of foliage or the presence of water can be smelt from a distance of five or six kilometres (3–4 miles). The trunk is also used for touching, and the hairs which surround its orifice are especially fine and sensitive. The nostril – a tube lined with mucous membrane – is contained inside, and has a highly developed sense of warmth. Everything which is to be taken up or sucked up is first of all very carefully sampled by the trunk.

The movements of this sense-instrument are so manifold and so expressive that its own sense of movement must be very highly developed. Georges Blond is quite right when he says:

> The trunk is so fully developed that it is possible that
> it informs the elephant of everything and transmits to
> it knowledge of the land even from a distance ... one
> seldom has the impression that the elephant is watching
> with its eyes. The trunk swings, turns, is raised, searches,
> seems to perceive everything.

We are therefore well justified in speaking of the sense of movement which perceives the forms and lines of the environment as having its seat in the trunk, and saying that the trunk largely replaces the eye in perception of space and its forms.[4] It acts as a sense-organ for receiving impressions of touch, pressure and warmth, and as a limb for gripping, beating and buffeting.

Moreover, the elephant's feelings and drives, emotions and moods are expressed through the trunk: they are revealed by the sounds which it produces and the gestures it makes. The trunk has at its disposal an array of sounds, ranging from a loving grunt to a frightening trumpeting which seems to shake the earth.

On a much lower level of the animal kingdom and anatomically different we find something which is of a similar nature in the realm of the cephalopods; the tentacles of the nautilus and the mighty arms of the giant squid and the octopus can be compared to the trunk. Here also we see the development of a huge 'lip' which has become a limb, and at the same time has extensive sensory functions. We have here a trunk-like formation within the realm of the invertebrates. But how different it is in gesture and nature. These tentacles are organs of voracity and the fearsome arms of the giant squid, which can be many feet long, grasp their prey with their suckers, and do not let loose. The trunk of an elephant, however, is an organ of sense, contact and grip. It provides the animal with a phenomenal spectrum of sentient experience that seem to reach the extremes of joy and distress. It is as though the inner life of the elephant is concentrated into its trunk.

Alongside the trunk the eyes almost disappear. Behind them are the massive ears, spreading out from the base of the trunk like great wings. The ears, too, indicate what is going on inside the animal: thus they reflect inner tension or relaxation. Something the trunk detects will make the ears prick up, while noises heard by the ears will alert the trunk and set it searching around. Both organs, ears and trunk, work together and complement each other in a remarkably complex fashion. Their range of sense-perceptions reaches out to embrace warmth and cold, sounds and scents, wind and weather and movement.[5]

The elephant's characteristics

Behind these powerful sense-organs lies a relatively small brain. The cranial cavity – much smaller than the huge skull – is suspended in a network of pneumatic bones that are filled with air and coated with mucous membrane. Thus the brain is hidden inside and is far removed from the outer walls of the skull. These bones are in direct communication with the nasal passages leading in from the trunk and can be compared to vastly enlarged sinus and facial cavities. This unusual arrangement makes it possible for scents and all impressions carried on the air to penetrate into the interior of the skull and to envelop the brain with messages from the atmosphere. The brain is in a sense embedded in the atmosphere and so becomes a kind of living, sensing barometer.

This is certainly relevant to something reported by many observers – that streams and rivers, sand and clay hills, and plains, are smelled or sniffed out from a kilometre or more away. The expanses of air and earth are an open book wherein the elephant does not read but smells, senses and perceives.

In order to maintain its huge bulk the elephant consumes and excretes an enormous quantity of food. The amounts which pass through the organism are scarcely credible: one reason is that the digestive process is very inefficient. Almost half of its purely vegetarian nutrition leaves the body undigested after passing through 35 metres (115 ft) of intestine. Elephants eat between 150 and 170 kg (330–375 lb) of food in 24 hours, drink 70 to 100 litres (18–26 US gal) and produce 80 to 110 kg (175–240 lb) of excrement per day.

No enclosed territory is large enough to satisfy this huge consumption of food. That is why elephants are wanderers, continually on the move, at home in forest and steppe, in savannah and mountain range, Richard Carrington writes:

> Wild elephants occupy a wide range of environments and seem equally at home in all of them. They adapt themselves quite happily to life in forest or bush, on high temperature plateaux or sweltering coastal plains, in the sultry valleys of Ceylon [Sri Lanka] or on the mist-covered slopes of African mountains between 8,000 and 12,000 feet [2,500–4,000 m] above sea-level.[6]

In terms of this astonishing capacity for adaptation, human beings and elephants are alike. The only difference is that in order to counter a bad climate people have to use all sorts of inventions, such as clothing and shelter, heating and air conditioning. Here we are behind the elephants, for the elephant masters the most varied environments through the astonishing adaptability of its inner world. Its body-temperature, too, is similar to human temperature (approximately 36°C) so that its body is like a pleasantly warm stove.

It took a long time to find out whether or for how long elephants sleep:

> Nobody had ever seen a wild elephant sleeping in the open. Never had a researcher or hunter been able to observe the sleep of a herd by moonlight or under a clear sky. An elephant's sleep is so short and above all so incredibly light that for a long time their only rest was thought to be a daily half-sleep, taken standing up.[7]

Benedict and Hediger made considerable efforts to resolve this question and both were eventually able to observe sleeping elephants. They had to take the most elaborate precautions for the animals are so hypersensitive that the presence of people disturbs them immediately. Both researchers were able to establish, independently of one another, that at about midnight adult elephants lie down for a short sleep of two to three hours, younger animals, however, sleep longer.

This short time should not puzzle us, for the brain of these giants is relatively small and the level of their consciousness fairly dull, centred chiefly on the lower senses. A dreamy picture-perception flows through the elephant and renders long periods of sleep unnecessary.

Here one can see the application of what Rudolf Steiner said:

> We can rightly observe the soul-life of the animal only when we eavesdrop on the self-enjoyment of its bodily nature. That is the essential thing. We observe only dimly the nature of an animal when we observe how it enjoys the external environment – but rightly when we observe how it experiences its own digestion. One must enter into the realm of the organs if one wishes to experience the epitome of the soul-experience of an animal. The animal is inwardly satisfied with this experience and anything added to it from the outer world is meaningful for the animal only so far as it can be lived out within its soul.[8]

If we apply these indications, some of the elephant's characteristics that we have noted will be clarified. This animal is surely a borderline case. The elephant seems to be less given over to its organs of digestion than are other animals, since it transforms only about half of its nutrients. It dreams less in the sphere of metabolism than does, for example, the cow, which transforms its food much more intensively. The rest is reserved for dreaming in the process of the senses. Animals experience that which 'can be lived out within its soul.' This in the elephant is a dull and quasi-clairvoyant perception of its environment, the realms of earth, water, and air. Not the 'all-penetrating light,' nor the brightness of day or the darkness of night, glittering with stars, are within its experience. For this reason its bodily exterior is equally grey and dull. No pattern enlivens its skin. Its inner life, too, is monotonously grey in terms of light; but in this twilight weave the clouds of scents and odours, the voices of wind and waters, the hot vapours of the valleys, and the storms blowing cool from the mountains. The rushing of rivers, the gurgling of springs, the cries of animals and the breaking and snapping of twigs and leaves under the myriad footfalls of animals. Just as light gives human beings thoughts, the elephant derives its peculiar wisdom from the endless variety of smell-perceptions. And these giants experience as noise what we hear as sounds and tones.

Humans understands something when the sound-impression of the word is taken up by the light of thinking. When in elephants thousandfold smells meet a multiplicity of sounds, a dull comprehension arises leading to action. The trunk – as we said just now – is not merely a nose but also a limb which sweeps around in front and acts as a guide. No concepts, no ideas and no aims arise. But impulses for action, instincts and urges, are present. They have been working in the species for millennia in the same way and they guide the herd and the individual animal from place to place.

'The animal brings to earth what it is capable of and what its being allows it to experience,' says Rudolf Steiner in the lecture quoted above. And he adds, 'What in its soul does the animal experience? It experiences its species from birth to death.'

This holds good for the elephant. It experiences the nature of its species and lives in the consciousness belonging to it. Immersed in this current and moving with it the elephant fulfils its task on earth.

A sense of community

Sometimes there is a lone elephant, usually a bull who, having grown old, has left the herd in a state of sickness and embitterment. Now completely wild, he lives by and for himself.

Occasionally two or three such elderly gentlemen will come together to form a pseudo-community apart from the rest.

Otherwise the elephant lives in the community of the herd. This consists of individuals more or less closely related to one another. Carrington says:

> The herd will include parents and children, brothers and sisters, uncles, aunts, nephews, nieces, and possibly a sprinkling of in-laws who have been accepted into the herd from outside. The number of individuals will vary considerably, ranging from ten or twelve in a small herd to fifty or more in a large one.[9]

From time to time the herds find themselves in a much larger community. This happens only when a shortage of food or water drives them together in particular places. Then they stay together during the time of shortage. Sometimes there will be some exchange of a few animals between the herds, but soon the family groups move away together as integral units.

It is not a question of family groups in the human sense. The elephant is not monogamous. The bull associates with a female during the mating period, stays with her for the first few months of pregnancy and then leaves her. Mostly it is the female elephant who loses interest in the male and devotes herself entirely to her future offspring. The pregnancy lasts for 22 months and usually only one baby is born. Rare cases of the birth of twins have been observed. The newborn elephant is very closely tied to the mother, who often chooses one of her comrades to assist her in bringing up her child.

A few hours after birth the young elephant, weighing about 100 kg (220 lb), is able to get up and walk, and in a few days becomes a member of the herd. It grows up with the other young elephants, plays, romps and indulges in plenty of mischief. Maturity is reached in eight to ten years. A female can become pregnant at an age between ten and fifteen, but the proper time for maturity is only after the twentieth year, as it is with humans.

The mothers are totally devoted to their calves. They look after them and protect them for years, and later, when the young are fully grown, they are still recognised by the mothers, probably by smell, as their own.[10] Usually if not always, the herd is led by a female. When the herd is on the move, the females tend to go in front, with the leader at their head. They keep the youngest among them; then come the young elephants, and last of all the young bulls. The community of the herd is a unity of soul. A sick or wounded member is taken into the midst of the others and cared for.

The herd is a close community of blood, which lives together, experiences things together, and responds to communal instincts. Young elephants show individual features which slowly vanish during growing up and reappear in old age. Then the aged ladies and gentlemen separate themselves from the herd and carry on their own lives.

In ageing, too, the elephant is similar to human beings. It remains capable of reproduction up to its sixtieth year. At this time the molar teeth stop their self-replacement, and with their disappearance the process of ageing spreads gradually over the body. Are there really elephants' graveyards to which the old animals go to die? Whether or not this is just a legend may never be known; too many elephants now die from persecution and massacre.

It has been possible, however, to piece together what their natural living conditions used to be – a family herd led by females, forming a community which extends beyond the individual animals. This is reminiscent of the earliest form of human society. Three other characteristics in which elephants have something in common with humans are their age of maturity, life-span and body temperature; also the fact that their tusks are permanent teeth which at an early age replace the milk teeth.

These various traits go to indicate that a close biological link exists between elephants and humans, although in bodily form they have drifted far apart. And so arises the question – where have elephants come from?

Origins and migrations

We have already mentioned the astonishing fact that in prehistoric times some 352 species and families of proboscides were distributed over almost all the continents. The last surviving descendants of this vast group of animals are the African elephant *(Loxodonta africana)* and the Indian elephant *(Elephas masimus)*.

Until the beginning of twentieth century their origins were a mystery. Several skeletons of extinct proboscides had been found. They were giants, some being a little larger and some a little smaller than present-day elephants. Their tusks either protruded directly forward, as in *Anancus* and *Palaeoloxodon*, or curved upwards as in the various forms of mammoths. The construction of the skull varied widely, and the bones from feet and legs showed many dissimilar forms. But these animals were branches on a tree whose stem had not yet come to light. Then, between 1901 and 1904, when two English researchers, Beadnall and Andrews, were examining the regions surrounding the old Lake Moeris in the Egyptian province of Faiyum, south west of Cairo, and searching for fossils, bones from hitherto unknown animals were discovered. Parts of an ancient sea-cow *(Eosiren)* were found, then a strange animal with a bony comb on its forehead *(Arisnotherium),* and three different types of primitive animals with trunks. They were grouped together in a family and named *Moeritherium,* after the place where they were found. These specimens, not much larger than present-day pigs, already show signs of tusks in the upper jaw. The skull also has the bone-cells containing air, such as we have noted in the elephant, and the bones from the limbs indicate a build which is found only in Proboscidea.[11]

Carrington, from whom I take these details, writes:

Much of northern Africa was covered by an ancient sea called Tethys and instead of the dry Sahara a vast region of swamps and fertile plains extended from halfway across the present Atlantic Ocean to the easternmost bounds of Arabia. This was the environment in which scientists believe that *Moeritherium* made its home, and several points of its anatomy suggest that it was a swamp-loving animal, partly aquatic in the habits.[12]

Thus we have a relatively small animal, possessing from the outset many marks and characteristics of the elephant, which began to develop in the waters and swamps of North Africa. In this same region researchers then found the remains of a similar animal the size of a tapir which appeared later but still co-existed with the *Moeritherium*. It is assumed that this larger animal was already more adapted to living on dry land. Living at the same time, however, were the sea-cows, which were totally water animals and yet close to the family of the Proboscidea

Thus we find the roots of the elephants and of the animals related to them in a world of waters and swamps. Their cradle was in the north of Africa and from there they spread north, south, east and west, all the time growing larger and more powerful. With increasing size the trunk developed as though to counterbalance the larger and stronger body.

Carrington believes that: 'The first migrations were to Asia which then became a centre for further development and dissemination. It was from this second home, as well as from Africa, that the Proboscidea undertook their widespread migrations over the earth.' It is also assumed that they reached America only via Siberia and Alaska.

Such concepts, however, can be held only if one overlooks the existence of the continent of Atlantis and its history. But if we take seriously the many indications which confirm for us its antediluvian existence, then we have a better understanding of the pre-history of the Proboscidea. In the early days of Atlantis they begin to appear in what is now North Africa. During the Eocene and Oligocene epochs the *Moeritherium, Eosiren* and *Phiomia* emerge.

From the Miocene onwards, the time probably corresponding to the first two epochs of Atlantis, the large and powerful Proboscidea appear as it were all at once – *Dinotherium* in the east, on the Asian continent, and the mastodon in the west, in the centre of the Atlantean continent. As a dominant species they spread across the world and prepared the ground for the human being as he gradually came to terms with the earth. They are the pioneers – only in their wake could man follow; they prepared the conditions of life for their human brothers.

Rudolf Steiner often said that atmospheric conditions during the Atlantis era were very different from those of our time. Air and water, still intermingled, enveloped the earth, and all existence was enveloped in perpetual mists. The contours of things swam in the thick veils of mist and neither sun nor moon nor stars could be perceived as we see them today.

The surface of the earth was soft; vegetation was rampant, and tropical-type plants covered the ground. If one goes by these indications, then only can one understand the real nature of the trunk in the Proboscidea. This organ was the only means or orientation in the mists. It could smell and detect, touch and perceive much that was still hidden from the eyes. The trunk was the periscope by means of which these powerful creatures found their way through the water-vapour and fire-mist of Atlantis.

Gradually, by eating and digesting they helped to overcome the excess of vegetation and so became pathfinders for the progress of humankind as it then was.

Today the elephants still love the watery element. They are the best swimmers among the land mammals and can cross the widest and swiftest rivers. They can also wade through shallow rivers with just the tips of their trunks showing. Thus the trunk has remained as a kind of periscope and is used as such from time to time. Carrington writes:

Elephants will sometimes go swimming, or wallow in muddy pools, for the sheer joy of being in the water. An elephant bathing party is a most amusing sight. The animals splash and trumpet, squirt water over themselves, or lie at full length with the contented expressions of elderly gentlemen surf bathing at the edge of the sea.[13]

Blond indicates this inner connection with the water:

At times the elephants in Chad spend half their lives in the water. In the antique teachings of the Berbers and the Neolithic civilisation of the Nile one finds again and again connections between elephants and water. Numerous drawings on cliffs portray elephants near waves, the symbol of the watery element.[14]

He closes this chapter, in which he describes a herd bathing, and coming out of the river at night, with the following words:

They loomed tall over the bank, enormous black shapes, gleaming like anthracite. They were an image of life in the beginning, born in the night, emerging from the water and on their way to the conquest of the earth.

Man and elephant

There must once have been an intimate connection in life between man and the Proboscidea. They conquered for man the space in which he later settled, having followed them there. The human tribes at that time were as closely blood-related as are the families of elephants. The dominating social form among men, too, was a matriarchal leadership; only by degrees did new social forms arise.[15] Then came the time when the continent of Atlantis was gradually submerged under catastrophic floods, and the present-day order of the earth's surface gradually took shape.

The doom of the Proboscidea was sealed by this cataclysm. Those which did not perish directly slowly died away because of the changed environment. Some, such as the northern mammoth, tried to adapt themselves to the ice age, but they were destroyed by the snow, the cold and the perpetual winter. These giants died in the struggle with yet more powerful elements; their time was past.

But along with the people who left the sinking continent and travelled eastwards there were some animals, and from them arose the later elephants. Rudolf Steiner distinguishes a northerly and a more southerly stream:

> We have here a stream in the development of humankind which has its origin in certain peoples moving from the old land of Atlantis to a more northerly region. As they move they touch on areas which today include England and northern France, Scandinavia, Russia, and stretch as far as Asia, into India. Another stream travels more to the south; its path runs from the Atlantic Ocean through southern Spain, through Africa and over to Egypt and from there to Arabia. These two streams, equally large migrations, pour forth from Atlantis towards the east.[16]

Here we find the explanation of the two families of elephants which still remain. The Indian elephant accompanied the more northerly stream, and the African elephant migrated with the more southerly stream. They survived because they attached themselves to people who were fleeing from Atlantis and seeking for new sources of life; they are monuments of a past age which have survived up to the present day. A whole series of cave and cliff paintings in France and Africa are witness to this interpretation. In the style typical of the Stone Age, which corresponds to the migration period of Atlantean man, the outline of the elephant is clearly depicted. The celebrated drawing from the

El Pindal Cave in Spain shows an elephant with the indication of a heart, and in northern France mammoths and elephants are portrayed with straight tusks. It may indeed be possible that the Indian elephant is a descendant of the *Anancus*, while the African elephant has developed from the mastodons.

From Greek and Roman times coins are extant which depict the elephant as a worshipper of the sun and moon – perhaps because it saw these heavenly bodies only after the submergence of Atlantis, when a clear atmosphere came into being, penetrated by light and stars.

It was especially in India that the elephant was revered and worshipped. The god Indra was portrayed riding on an elephant named Airavata, representing the archetypal elephant, from whom all others are descended. Rudolf Steiner, in the lectures cited above, describes Indra as the mightiest of the gods at work behind the phenomena of the atmosphere, and at the same time engaged in the whole process of our breathing. If we consider the intimate relationship between the elephant and the atmosphere, as described earlier, the image of Indra riding upon Airavata acquires a new and special meaning.

Until recently, the highest honour was accorded in India to the white elephants, possibly albinos. If one was captured, it was led in a festive procession to the throne of the ruler, received by him with deepest reverence and taken to a specially erected building. Attended by servants and fed with the best food, it spent the rest of its life there. There are some addresses which were given at the reception ceremony of a white elephant. The conclusion of one of them runs:

> O father elephant, we beseech you, give up your desire
> to remain in the forest. Look upon this wondrous
> palace, this heavenly town! It overflows with riches and
> everything your eyes could wish to behold or your heart
> could wish to possess. It was your own merit which
> brought you here to see this town, to delight in its

richness, and to be the favoured guest of His Majesty, King of Kings.

In the ritual songs of the African pygmies, intoned after the elephant hunt, they speak of father elephant:

> Our spears went astray,
> O father elephant!
> We did not wish to see you dead,
> We did not wish to make you ill,
> O father elephant!

This identification of the elephant as a father-figure occurs repeatedly in the ritual songs and sayings of India and Africa.

The white elephant is like an inkling and premonition of a messiah. The people who meet it experience a kind of illumination. Out of the dark skin came the light, shining apparel, bearing a light of the future.

From the sphere of the creation the fatherly Airavata came forth as the servant of his lord Indra. Towards the end of ancient Atlantean times the earth was pervaded with his creations; they helped to clear the atmosphere and to lighten the ground. After their work was done they were called back, and only some witnesses of their old power remain.

Those who were once kings of the earth are now beggars and wait for their redemption. Through them the groans of creation are to be heard with great clarity. They who were once lords had to become servants, and they willingly give themselves to their destiny.

5. The Bear Tribe and its Myth

Animals in myth

Individual animal groups each have their own specific identity that appears, thrives, becomes widespread, and finally to disappear again from the realm of earthly forms. Thus they have their own special and quite personal destinies. We should never let ourselves be led astray by considerations of an animal's usefulness or harmfulness, although we find this approach again and again in the writings of Brehm and his contemporaries as well as in works by modern writers; they are thinking in terms of a threadbare teleology based on human predominance. An object can be useful, but an animal is a being. Neither can human beings be judged as to whether they are useful or harmful; every human being is worthy to live on earth, or he would not be born.

In the same way, each animal group should be seen in the light of its own worth. Although animals always occur as members of their species, family, group and class, they have a task in the plan of creation which only they can carry out. To recognise and to describe this task would be the goal of a future zoology. Current behavioural research is a first door opening on this new vista. When ethnological phenomena receive a more spiritual interpretation, a further door will open. For this, however, the relationship between man and animal that has developed in the civilised countries of the world must not be taken as the standard.

In large and small cities, housing developments and suburbs,

people live almost totally apart from the world of animals. They have created a 'reservation' for themselves and left the remaining bits of wood and meadow to the animals. Where there are still natives untouched by today's civilisation – in the forests and swamps of the Amazon, in certain parts of central and southern Africa, in Borneo, the Celebes and New Guinea – here human beings and animals live in much closer community; they are aware of one another and behave accordingly – with, against, and for each other, forming a living unity, a symbiosis. The animals have a deep effect on the lives of the human beings, who in turn have a critical effect on the existence of the animals. It is not just fear and superstition which lie at the root of taboos, festivals and magical practices. The soul-world of the animals – their deeds, their behaviour, the fantasies and supersensible experiences connected with them – permeates the picture-world, the feeling and doing of the natives living with them.

An earlier humankind was even more closely related to the animals. There was a time – during the Atlantean period, for example, the last traces of which were still known to Plato – when not only gods and heroes walked in human form among the awakening sons of the earth, but when the group-souls of the animals were also near to man. They had effects in the realm of the sense-world, and produced their own deeds and achievements.

Gods, heroes, animal-souls, human beings and animals shared life with one another. They communicated in deeds and cults, in rituals and ceremonies; they touched one another and mutually influenced one another's destinies.

The residue of these events we find in the bones and fossils which come to light as the remains of bygone cultures. However, we should not assume that only these residues existed formerly; much of what was living then could leave no traces: its material basis was too rarified and fragile.

Bodies and forms were far more amorphous, more changeable, than they have become today. Moreover, the division between animals and humans still had many bridges. Morphological transitions existed not only to the apes, but also to the carnivores, the ruminants, the elephants and walruses, the seals and horses. Pans and centaurs are not fantastic chimeras of the primitive folk-mind: they are the bridges that once linked the evolving animal groups and the developing human races with one another. The archaeopteryx is a transitional form altogether comparable to the centaur: it links bird and reptile, while the centaur bridges the gap between hoofed animal and the human being.

The magical power in fairy-tales, by which people could be transformed into animals and animals into humans is a relic of natural events that once occurred continually. Transformation was a primordial mode of behaviour of all living things. Its power once extended even to the physical form; today it is limited to our fantasy. In our dreaming and imaginative pictures, fantasy is still capable of creating what once was physical, bodily reality.

The bear family

Where the constellation of the Great Bear looks on the earth by night – the Arctic Circle, the forests of Canada and Siberia, the northern swamps of Russia and many provinces of Scandinavia – there the bear tribe is at home.

It extends further south as well, but with a reduction in range and numbers as it approaches the Tropic of Cancer. The forms of the several families grow smaller and lose in dignity. Two thin arms stretch down across the equator in south-east Asia and in South America.

A mighty and powerful race inhabits the northern hemisphere, dwindling and thinning out towards the south. The various groups in the family lie almost like belts around the earth, varying in form as they occur in Asia or in America. They appear to be the base-circles of many cones, all having a common tip: the Pole Star and the Great Bear which circles it.

In the icy wastes north of the Arctic Circle, lives the *polar bear*. He is at home all around the Pole. In Brehm we read:

> On the eastern coast of northern North America, in
> Baffin Bay and Hudson Bay, in Greenland and Labrador,
> on Spitsbergen and other islands, the polar bear can be
> seen both on solid land and on floating ice. On the island
> of Novaya Zemlya, in northern Siberia, wherever there is
> Arctic ice, there too the polar bear lives.[1]

A second ring, stretching from eastern Asia to North America, is formed by the *black bears*. They then spread southwards over Canada and regions of the United States all the way down to Mexico.

The *brown bear* is the ordinary, widespread bear, known to all. He attains various sizes, has differing colours and forms and zoologists do not have an easy time keeping the various subgroupings apart. In Brehm's words:

> If we unite all the forms mentioned into a single species,
> then we have an area of distribution that stretches from
> Spain to Kamchatka and from Lapland and Siberia all the
> way to Lebanon and the western Himalaya.

In Europe brown bears still inhabit almost all high mountain ranges: the southern Alps, Abruzzi, Pyrenees, Carpathians, the Balkans, the Caucasus and the Urals. They dwell in the forests of Sweden, Finland and Russia; they populate northern and Central Asia, and can be found in Syria, Palestine, Persia and Afghanistan. Particularly grand specimens live in north-east Asia, around the Amur Basin, on Sakhalin and Kamchatka (the Kamchatka bear). The North American brown bears and grizzlies are also most imposing figures; and the Kodiac bear of Alaska is the largest carnivorous animal in the world.

In southern Asia, the dense forests and swamps of Burma are home for the *Asiatic black bear,* which can be seen also in northern

India and in Kashmir on the slopes of the Himalayas, and has pushed north from there into south-east Siberia.

Still further south live the *sloth bears*. They are distributed across the whole of India and are known also in Sri Lanka.[2] Towards the south-east, the Malayan *sun-bear* extends to Sumatra and Borneo.

The two groups which reach across the equator are the *spectacled (Andean) bear*, inhabiting the Cordilleras as far as Bolivia, and the Malayan sun-bear. In south China and America live the small bears, only distant relatives of the true bear, with marten- and cat-like forms (raccoon, honey bear, coati, lesser pandas), but one has a typical ursine form, the black and white giant panda.

The distribution of the bear family over the earth thus shows clearly that the northern zones are its home. There are no bears in Africa, or in Australia and New Zealand and Antarctica. The bear belongs to the destiny of the three northern continents, where it has taken deep root in human life, action and intuitions. The human being is attached to the bears; he respects, fears and honours them. He fights with them and tries to overcome them.

Variety and mutability

Bears are a sub-group of the carnivores. They form a closed family in their own right, like the dogs and the cats. Zoologists place the bear near to the dog. There is also some evidence of an origin in the European north but it is extremely difficult to speak of this with certainty. In any case, bears are closely connected with the coming and going of the ice-ages. In many European caves great accumulations of remains of the extinct cave-bear (*Ursus spelaeus*) are found. The remains of these giant bears, the most frequent quarry of stone-age hunters, strike us by the variability of head and face formation. Othenio Abel says:

The skull forms of the cave-bear from these layers (later ice-ages) vary quite widely, from types with altogether flat foreheads, reminiscent of the primitive ursines, to forms with exceedingly steep brows ... Some skulls stand out by the length of the muzzle, while in others the face is remarkably shortened ... Otto Antonius quite rightly pointed out that the range of variation in the Mixwitzer cave-bear skulls reminds one strongly of certain kinds of dog (German Shepherd, Great Dane, pug) ... Doubtless what we see here is simply an unusually great variability in this bear, surpassing by far the range of variations visible in the Kamchatka bear.[3]

All bears display this unique ability to vary the basic form of their type in many ways. Thus Brehm writes:

Among the bears, not only the colour but also the skull seems extraordinarily variable. Indeed, individuals from one region are usually not hard to distinguish by colour from those of a far-distant place. However, if one looks at the forms occurring between these two, and at the whole range of variation, then distinction of the separate forms becomes difficult.

This indication is of great significance; for it characterises the entire race of bears. Apart from the southern types, the bears have no properly definable or distinguishable species; even the polar and brown bear are so close to one another that crossing leads to fertile offspring. They are as mutable as the breeds of dogs. Among dogs, this flexibility is connected with their nearness to human beings. The bear, also, is very close to the human being. Even today certain tribes, such as the Ainu in northern Japan and on the Kurile Islands and Sakhalin, as well as the Tungus and Gilyaks, have a most intimate, in fact, religious relation to the bears that live around them.

What humankind has been able to bring about with dogs, through taming and training, they have never succeeded in achieving with bears. Here and there, of course, there is a dancing bear, but the race as a whole has remained wild and free. The bear is close to humans; but it appears to be of equal birth, and has never let itself be subjugated. Bears flee from people who would destroy them, and leave in peace those who are well-disposed. The bear can live with humans so long as it is respected as an independent race in its own right. The range of variation which the race of bears has preserved in the form and colour of its body, reminds us of the great racial variability of humankind. Here, there are connections which cannot be fully grasped today; but bear and dog both show this characteristic, which is not encountered in other creatures.

The nature of the bear

Bears not only live near people; in quite a few other respects they are more closely related to us than most other animals. For one thing, they walk primarily on their sole, setting the entire foot on the earth, including the heel. The sole, except in the polar bear, remains hairless, and is cast off once a year.

The bear can stand erect; he comes at his opponent in upright posture, and confronts his foe erect. In Heinrich von Kleist's marvellous piece *On the Marionette Theatre,* such a battle is described:

> The bear, as I stepped amazed before him, stood on
> his hind legs, his back leaning against a pillar to which
> he was attached, his right paw raised in readiness, and
> looked me in the eye: this was his fencing position.

And after the fencer has exhausted himself in vain efforts to strike the bear:

It was not only that the bear, like the foremost fencer in
the world, parried all my thrusts; to feints (and no fencer
in the world will match him here) he made not the
slightest reaction: eye to eye, as if he could read my soul,
he stood with his paw raised ready to strike and when my
thrusts were not in earnest, he did not move.

When the bear takes flight, he turns four-footed and moves
away with his ambling gait. Poppelbaum calls the bear's gait
'shuffling and a bit sloppy, a shambling trot; he has no true gallop.'[4] This mode of locomotion is connected with the fact that
the bear's front limbs are shorter than his hind legs. They are
arms, not front feet; with them he can stroke off ripe berries
from bushes and bring them to his mouth. He can dig up anthills
and put the contents, larvae and mature ants into his open jaws;
likewise he can open honeycombs, take out the honey with his
hand and smear it into his mouth.

A bear skeleton mounted on all fours gives the immediate impression that this animal has sunk over forwards. The
backbone, with the head appended to it, falls forward over the
short arms. It would be wrong to think that the bear was once
four-footed and later stood up erect. It was the other way round!
The bear was a creature that once went about upright, but then
slumped over forward, just as he does today when he flees or
roams. It was a plunge into animality that he once took.[5]

All observers tell of the bear's double nature. On the one
hand, they speak of his harmless, good-natured disposition, his
peaceable attitude toward people and the patience he usually
displays. Krementz, a forester, sums up his long experience
with bears in the Pinsk Marshes in southern Belarus in these
words:

I have not heard of a single case where, during a bear's
roamings and encounters with men, he has attacked
them. On the contrary, in most such cases he flees ...

> The bear has a good nature, although he should not be
> trusted under all circumstances; in particular he does not
> like to be irritated, or taken by surprise while resting.[6]

However, once he is disturbed, startled, surprised or suddenly
attacked, the other side of his nature comes out. Then he stands
upright, turning wild and dangerous. Then he slashes animals
and humans, breaks into the pens and kills the cattle. So the
four-legged, ambling, good-natured plant-eater becomes a two-
legged, wild predator. Then he raises himself upright for battle.
But he is no real carnivore; as a peaceable vegetarian he comes
close to the tamed condition of a dog. When, however, he stands
up on two feet to fight, he grows beyond his carnivore-nature
and becomes a raging pre-human.

> It is a sign of his high intelligence that the bear regulates
> his behaviour quite according to the circumstances: he
> will be trusting or very careful, depending on what he has
> experienced. Though he may dwell in the vicinity of a
> village, for years he will never be seen by anyone, so well
> does he know how to hide himself and how secretively to
> go out for food at night![7]

However, the bear is not an expressly nocturnal animal. He
can be met almost anywhere by day, in lonely clearings in high
timber-woods, in dense marshy regions and in the jungle of
southern Asia. The bear avoids people; but I doubt whether he
fears them. He avoids them out of a sort of shame, which at
the same time is a form of nobility. He senses his deep kinship
with human beings. He has an intimation that he too was once
upright, as man still is today.

The bear, like man, has arms and walks on the soles of his feet,
but head and face are overgrown with a thick fur which has covered the arms with 'sleeves' and the hands with 'gloves.' Beneath
this covering, however – as in the story of Snow White and Rose

Red – lives the prince, whose golden garment sometimes shines through.

Is it so astonishing then that bears, born blind and naked and tiny as rats, need six full years before they are fairly well grown? This entire period they spend near to their mother; the father pays them no attention. These young ones are in many ways still closer to humans than are their elders. According to Gerlach:

> What is so comical to us about the young bears is their similarity to humans in many movements, in their weighty sole-gait, their ability to stand upright and to use their front paws as hands.

In the Walt Disney film, *White Wilderness,* we can observe polar bear cubs at play. They roll snowballs, throw them at each other, and slide like children down snow banks on their backs. They roll, tease and play as otherwise only human children do.

Is this, perhaps, why the teddy-bear has become a playmate so beloved by human children? Because the bear was never tamed, we have copied him and thus made him a companion for our children. The teddy-bear is the 'enchanted' form of the wild bear.

Human attitudes to the bear

This closeness to human beings, which emerges so plainly in the character of the bears, becomes still more vivid to us when we investigate human modes of behaviour towards the bear tribe.

The Ainu, whom we have mentioned already, have a particularly intimate connection with bears. In every village they set up special cages for raising bear cubs. The cubs taken from their bear-mothers quite early are suckled by Ainu women; later they grow up in the huts along with their foster-parents' children, until they are big enough to go into the cage. Here they stay for two of three years, until the bear-eating festival.

Frazer devotes a separate chapter to this festival, and attempts to bring together all existing descriptions. It is a yearly festival celebrated either in the middle of winter or in the autumn. The entire population of the village takes part. First, the father of the festival steps before the cage and addresses the captive bear:

> O thou divine one ... thou wast sent into the world for us to hunt. O thou precious little divinity, we worship thee; pray hear our prayer. We have nourished thee and brought thee up with a deal of pains and trouble, all because we love thee so. Now as thou hast grown big, we are about to send thee to thy father and mother. When thou comest to them, please speak well of us and tell them how kind we have been; please come to us again and we will sacrifice thee.[8]

Thereupon, the bear is brought out of the cage and led, bound, in a communal procession through the village. After this blunt arrows are shot at the sacrificial animal; and after it has become enraged, its neck is crushed between two strong posts.

Among the Gilyaks, the bear is led into each house in the village and fed porridge by each of the inhabitants. Afterwards it is pierced through the heart with an arrow, while the women cry and wail a death-song.

Among both Gilyaks and Ainu, the head of the animal, with the pelt still hanging from it, is installed in a seat of honour indoors, and its own flesh and blood are ceremonially offered it to eat. After this the entrails and flesh are cooked in ritual manner and consumed in a festive communal meal. The cup offered to the sacrificial animal is passed round, and everyone present must drink from it. This cup is called the 'cup of offering.'

The festive meal often lasts for several days. The women perform sacred dances and finally each person present, on stepping out through the door into the open, is given a light stroke with a birch rod by the village elder.

The Nanai call these sacrificial bears their son and brother; and many of the Ainu say that they are descended from bears, calling themselves 'children of the bears.' The bear, however, is god of the mountains; and so the people say: 'As for me, I am a child of the god of the mountains. I am descended from the divine one who rules in the mountains.'

Frazer attempts to understand all these isolated facts:

> We are expressly told that the Ainu of Sakhalin do not consider the bear to be a god but only a messenger to the gods, and the message with which they charge the animal at its death bears out that statement. Apparently the Gilyaks also look on the bear in the light of an envoy dispatched with presents to the Lord of the Mountain, on whom the welfare of the people depends. At the same time they treat the animal as a being of a higher order than man, in fact as a minor deity, whose presence in the village, as long as he is kept and fed, diffuses blessings, especially by keeping at bay the swarms of evil spirits who are constantly lying in wait for people, stealing their goods and destroying their bodies by sickness and disease.[9]

The individual bear kept and then eaten as a sacrificial animal is certainly a 'messenger' of the gods; for he is only a part of that great being, 'bear,' of whom these people have knowledge, and whose workings they recognise. Just as the human soul returns after death to the higher worlds, so the soul of the sacrificial animal is reunited with its 'father and mother,' that is, it enters into its spiritual home.

The Ainu, the Tungus, and the Gilyaks live surrounded by bears; and every night, especially in winter, they see the 'Great Bear' twinkling almost directly overhead. Quite probably they have an intuitive feeling that there, up above, lies the home of those beings who take on bear-form among them below. It is highly interesting to hear that:

111

The seven great stars that constitute the Great Bear have
not the faintest resemblance to the form of a bear, yet
they appear to be known almost universally by this name,
even among savages in whose country no great bears
exist.[10]

Thus it is not an external similarity that prompts 'primitive'
people to give certain names to the constellations. Primitive man
is gifted in dreams and lives in pictures; he perceives the countenance of the bear-being in that heavenly chariot, and names the
constellation after this inner picture. Bayley adds:

This constellation was once known as the 'Sheepfold';
and it would appear from the emblems herewith that the
Great Bear was regarded as a symbol of the Great Spirit,
the Triple Perfection, the Light of the World, the Alpha
and Omega, or Jesus Christ.

Does this help us understand the image of the sheepfold, as
it is used in the Gospel of John? And the words of the Christ:

Truly, truly, I say to you, he who does not enter the
sheepfold by the door but climbs in by another way, that
man is a thief and a robber; but he who enters by the
door is the shepherd of the sheep. To him the gatekeeper
opens; the sheep hear his voice, and he calls his own
sheep by name, and leads them out. (John 10:1–3)

It could well be that the bear is the gatekeeper of the sheepfold. Or is he the thief and robber? In the First Book of Samuel,
David says to Saul, in order to get permission for his battle
against Goliath:

'Your servant [David] used to keep sheep for his father,
and when there came a lion, or a bear, and took a lamb

from the flock, I went after him and smote him and
delivered it out of his mouth; and if he arose against me,
I caught him by his beard, and smote him and killed him
... The LORD who delivered me from the paw of the lion
and from the paw of the bear, will deliver me from the
hand of this Philistine.' (1Sam.17:34–37).

Thus we come upon questions and hints. The city of Berne
keeps a bear-pen at the entrance to the old city, just as the Ainu
and the Gilyaks have their bear cages. Moreover, if we consider
the image of the cage where the witch keeps Hansel in the fairy-
tale, in order to fatten him for slaughter, is it not reminiscent of
these bear-pens in Japan and Switzerland?

Many cities (including Berlin, Berne and Bernburg) have a
bear not only in their name, but also on their coat of arms. The
bear is felt to be the protector of the human community settled
there; it is part of that Great Bear who circles the North Star and
watches over the axis, the gate to the earth-world.

Classical images

Thus images emerge and tend gradually to fit into a whole. The
Huzuls, for example, who live in the Carpathians, have such
respect for the bear that they will not call him by his name. He
is called 'little uncle,' or, most reverently, 'the great one.' The
Lapps, when they have hunted down and killed a bear, must follow a whole set of taboos. For three days the hunters undergo
certain purification ceremonies and only at the end of this time
are they permitted to step within the women's huts. Moreover,
the reindeer who drew the sleigh on which the dead bear was
brought in may not be touched or fed by women for a whole
year.[11]

The companions of the goddess Artemis were called the *arktoi*
– the she-bears; and Kerényi adds:

Artemis herself must at one time have been supposed to be a bear – or in more ancient times, when the fauna of Greece was more southern, a lioness.[12]

The story is also told that Callisto, a friend of Artemis, became pregnant by Zeus, who appeared to her in the form of a bear. However, like all the companions of Artemis, she had sworn eternal virginity and when the goddess perceived her friend's swollen body while bathing, she grew angry and transformed her into a bear. In this form she gave birth to a son, Arkas. Callisto, however, remained a bear and lived in the dark woods. It happened one day that she met her son, who in the meantime had become a great hunter. In Schadewaldt's version:

> The mother recognises him and stops, wishing to approach her son tenderly, but he sees in her only the wild animal, raises his weapon to strike and would certainly have killed her – but suddenly Zeus removes the two of them and sets them as neighbouring stars on the most beautiful spot in heaven, close to the heavenly pole.[13]

Are we not all like Arkas, who sees in the bear only the predator, and not the mother who gave him life?

The word 'bear' has a motherly, protective element in it through the letter *B*. Fuhrmann, perhaps going rather far, says:

> Intimately related to the series *bär* (bear), *boar, björn,* is the Swedish *barn* – meaning simply 'child.' Since the North Pole is designated and thought of as the source of life in such cases, it can be called the mother-point, and the first animal to come from it can well be called 'child' ... *Bär* is also *ab-ra* – that is, the absent sun and thus the north.[14]

It is not easy to judge whether or not such interpretations are justified. They give certain indications, but should be used only with great care. Still it is interesting to find the same linguistic root in the Nordic words for bear and child *(barne)*. Also connected with it are the words 'birth' and 'to bear', meaning to 'carry' or to 'tolerate.'

The word *Arkas* seems to be more important yet. Arkas, the son of the bear Callisto, appears in the older Greek sagas as the son of Artemis herself, when she still had the form of a bear. Arkas was the father of the people which inhabited the countryside of Arcadia in the Peloponnese. The Greeks called this folk the 'acorn-eaters,' and believed they were 'older than the moon'.[15]

'*Arkas,*' says Kerényi, 'is connected with bear, *arktos*.' Beyond this, however, the name has an obvious similarity to the name borne by one of the last great peoples of old Atlantis. They were called Akkadians, and lived in the northern regions of the later-submerged continent. Here the Akkadians met with the remnants of the primeval Hyperboreans. The latter were a human race that had hardly become earthly, living entirely in the realm of the soul. The Akkadians inhabited those areas which later – after the sinking of old Atlantis, that is at the end of the ice ages – re-emerged as the land masses of northernmost Canada, as Greenland, Scandinavia and Siberia. It is these regions within the Arctic Circle that we must look upon as the cradle of the bear tribe.

Is it not significant that in Greek mythology, Apollo, brother to Artemis, returned every year to the Hyperborean land? Moreover Leto, mother of both, came from that land. Apollo, in his being and his deeds, embodies the basic character of Akkadian man, whom Rudolf Steiner describes as follows:

> The men of the sixth subrace, the Akkadians, developed the faculty of thought even further ... The calculating faculty of thought sought the new as such; it spurred men to enterprises and new foundations. The Akkadians were therefore an enterprising people with an inclination to colonization.[16]

115

It may be that this urge to colonise also led them to the Peloponnese and gave rise to the Arcadians. Arcadia was also the home of the god Pan, whom Euripides calls the twin brother of Arkas. Pan was a sort of in-between being, like the bear. He, too, was good-natured and harmless, but could grow malicious and savage when disturbed while resting, particularly around noon. Pan, who appeared in many forms – as the son of Kronos, of Zeus, of Hermes – had goat's feet and a wild, bearded face.

The connections presented here are not meant as interpretations, but only as pointers in a direction where still more threads are to be found. Then it may be possible to grasp these indications in a more concrete and more relevant way.

A spiritual geology of the bear

Two characteristics of the bear will now be given particular consideration, for they point more clearly than any other to the bear's close relationship to man.

The bear is not a herd animal. It moves alone through woods and marshlands and it is rather a coincidence if several bears meet at a raspberry patch or by an especially bountiful salmon-river. Thus, in contrast to other carnivores (lions, wolves), the bear is a solitary creature. Nor does the male keep near his family; only in summer, during the mating season, do we encounter him with a female bear. Later, he leaves her; and in autumn she searches alone for a cave, a hollow tree-stump or a thick roof of leaves where she can give birth to her young in December or January.

No teleology or fantasies about adaptation can find a reasonable explanation of why the tiny bear-cubs come helpless and naked into the world precisely in the coldest season. This would be just the opposite of a 'struggle for survival,' for the father bear is also absent at the time, having retired to some cave-shelter for his winter's rest.

Rudolf Steiner indicated that in primordial times of human evolution – during the Atlantean period, for example – all human births took place around the winter solstice. He often said that in earlier times of the world's evolution, human reproduction was still bound to the course of the seasons. Conception could occur only in the spring, when the necessary forces became active, and thus birth could take place only towards the end of the year.

This behaviour changed, for humankind gradually emerged out of this nature-bound condition. Only among certain Germanic tribes, the Ingaevones for instance, did this trait persist into the third millennium BC.[17] The bears, too, have kept this trait. They still bear their young at the time of the winter solstice, and thus preserve a pre-human condition which has nothing to do with adaptation, but represents a remembrance of earlier times.

Are not the Huzuls right when they call the bear 'little uncle' – that is, the brother of their father? Or when the Ainu and Gilyaks hold their bear-festivals and call the sacrificial animal the herald of their patriarch? By holding the bear-feast they connect themselves with their ancestors. What they are performing is a pre-Christian mass, and it is quite probable that these Ainu are the last remnants of the former Akkadians. Their origin remains unknown. At first they were looked on as the aboriginal inhabitants of Japan but this hypothesis cannot be maintained, as Bernatzik explains:

> The most recent studies show that the Ainu probably played a very small part in the development of the Japanese people and its culture (Arctic elements) ... Particular bodily features of the Ainu are: strong, squat build, heavy hair-formation (especially head and facial hair), broad face, deep-set eyes with Mongolian eye-folds, and wide flat nose ... Their language must at present be regarded as isolated.[18]

Thus we have here a small racial group that stands alone, not showing Mongolian, Malayan or Indo-European features. These people are small of stature and heavily haired and they are intimately related to the bear tribe, as their festivals show.

We are reminded of similar elements in several German fairy-tales. In the story of Bear-Skin, a man turns for seven years into a kind of bear through a pact he makes with the devil. 'Hair covered almost his entire face, his beard was like a piece of coarse felt, his fingers had claws.' Also in the story of Snow White and Rose Red, a human being is transformed into a bear until, through the death of the dwarf who enchanted him, he recovers his human form.

So we have all these pieces of evidence, some indicating more plainly, others less plainly, that the race of bears was once a human group. Probably it was related to the Akkadians who lived in the far north, on old Atlantis, at a time when the ice ages began to set in. Present-day paleontology puts the origin of the bears (as we have mentioned) at the end of the Tertiary, around the Miocene. This geological epoch corresponds to the last phases of Atlantis. Von Gleich writes:

> During the second half of the Atlantean epoch, the warm, moist climate of the Eocene and Oligocene periods of the Tertiary gradually passed over into a cool climate, and finally to a very cold one. Of the Miocene epoch, Theodor Arlt says: 'In Europe, the palms recede southwards across the alpine region. Even in central Europe the leaves were affected by frost, despite the mild and damp climate prevalent there. Particularly in the Arctic the temperatures fell significantly. Thus it was the time when the North Pole shifted from the Bering Strait region towards Greenland, causing certain cold waves to strike Europe and so heralding the glaciation that was to come much later.'[19]

This was the time when some of the Akkadians migrated to the south, east, and west. Others, not feeling the unrest which the awakening power of thinking gave to their kin, began to make ready for the oncoming winter of the world. These were the incipient bear family. All around the Arctic Circle they began to spread, and the descending ice-masses brought them south to roam across Europe, Asia and America down to the Tropic of Cancer. Towards the North Pole they became polar bears; further south they developed into brown bear, grizzly, and all the other kinds we know.

The bear is the animal of winter. He has covered himself with a thick coat of fur and withdraws to his caves from the onslaught of snow and cold. In the form of the polar bear, however, he is able to take up the battle with winter. That is why bears love honey, the precious gift of the summer sun, for through the forces active in it, they can withstand the cold. The wonderful forty-sixth rune of the *Kalevala* speaks of the bear:

'Tell us of the birth of Otso!
Was he born within a manger,
Was he nurtured in the bath-room.
Was his origin ignoble?'
This is Wainamoinen's answer:
'Otso was not born a beggar,
Was not born among the rushes.
Was not cradled in a manger;
Honey-paw was born in ether.
In the regions of the Moon-land.
On the shoulders of Otava,
With the daughters of creation.'

High in the crystal-clear winter-heavens of the north, on the shoulders of the Great Bear – that is where the bear is born. He belongs with winter-nature; he is the winter carnivore who once was human, and still bears this humanness hidden within him.

His companions went south. There they gave rise to the figures of Pan, Silenus and the satyrs. Artemis and Apollo, to whom the bear once belonged, told the fate of the bears to the Greeks and to all following generations of humankind. It is the fate of the ice-ages, the fate of the sinking of Atlantis, the fate of humanity sinking into animality. And so we can understand Rudolf Steiner's admonition:

> It is not in the geological strata of the earth that we have
> to delve if we wish to know the man of prehistoric times,
> the man whose higher corporeality was still outside the
> physical body. To burrow in the earth is quite absurd;
> in the earth we shall never find traces of prehistoric
> man which are anything but decadent. But in the strata
> of human spiritual life, in the strata of spiritual geology
> which have been preserved for us in the wonderful
> Greek mythology there we shall find the normal, average,
> Atlantean man, just as in the geological strata of the earth
> we find snail shells and mussel shells. Let us study the
> configuration of the fauns, of Pan and of Silenus; it is
> there that we have the spiritual fossils which lead us to
> the earth's prehistoric humanity.[20]

We have attempted to trace this 'spiritual geology' of the bear and now we have an intimation of its origin. It still has a while to wander before the words of St Paul (Rom.8:22), that the whole creation groans and thirsts for salvation, are fulfilled for the bears. The Kalevala actually refers to the bears' pilgrimage:

> 'Otso, thou my Honey-eater,
> Thou my Fur-ball of the woodlands,
> Onward, onward, must thou journey
> From they low and lonely dwelling.'

6. Swans and Storks

The world of the birds

Even as children we noticed that spring was coming when the birds began to arrive from the south. Outside the sun rose ever higher, and deep in the soul joy arose. The birds as they returned brought back the growing light to us children. Spring and the awakening of the bird world belonged together. The fact that other animals reappeared was – apart from the butterflies, which heralded summer – not so important.

For a long time the return of the birds and of the sun must have been bound together for the human soul: it was as though the intensifying light and warmth embraced the bird world within them. Just as fish belong to the water, so do birds belong to the life of the air, pervaded by light and warmth. That is their kingdom. Here too, as everywhere in the realm of creation, there are exceptions, but they serve only to prove the rule.

In this realm of air, light, and warmth birds develop the special activity – flying – for which they are equipped. Usually a bird does not have to learn to fly; it flies naturally. A chick is at first unable to get itself into the air, not because it cannot fly but because its body is too heavy for its undeveloped wings. A growing bird never has to learn flying from its parents, unlike the young seal, who is coaxed for weeks by the mother to get into the water and attempt to swim.

A bird flies, as fish swim; its feathers carry it through the air. It has only to give itself up to them to progress forwards. It is this submission to the feathers which makes flying a matter of course

for the bird. The plumage is spread across its body like a wonderful garment which carries it over lands and seas, as flying carpets do in fairy-tales. What the bird has to do is to make itself part of the warmth and the air currents which flow across the earth so that they carry it along. It experiences the wind as it blows, the weaving of the warmth, the shimmering of the heat: its being is united with them.

The muscles of a bird's breast and limbs are astonishingly delicate in contrast to the power they would have to produce if it were only muscle-power which enables a bird to fly. The wings are needed for take-off and landing; flying is achieved by the feathers. They are a mysterious structure, formed only after completion of the main phases of the development of the embryo. They are added to the bird's body as it were from the outside, giving it form and structure, and gradually clothing the naked miserable body in beauty and dignity. The plumage is like a coat woven from powers which do not belong to earth and water, but to air, warmth, and light.

Thus we can understand Rudolf Steiner's saying that the bird experiences its bones and organs somewhat as we experience burdens such as suitcases or rucksacks we have to carry. 'You would not call this luggage your body. In the same way the bird, in speaking of itself, would only speak of the warmth-imbued air, and of everything else as the luggage which it bears about with it in earthly existence.'[1]

The bird identifies itself solely with the air which it breathes in and warms through. Everything else is foreign to it, not its own, but its burden. But this 'luggage', the bird's body, is constructed in a special way.

A bird's organs are all packed together in a small space; one might almost say that they are stowed in the 'rucksack' between the breast and the abdomen. Here are the heart and stomach, lungs and intestines, kidneys and reproductive organs. This space is almost completely enclosed by the ribs and sealed off at the front by a powerful breastbone. There is no bladder, so that urine and excrement are passed through a single channel. Neither is there a colon, and the breast is not separated from the abdomen by a diaphragm.

By contrast, the head is extended on a neck which may often look too long, or is sometimes quite short. The head is not really a head in the strict sense, but more like an appendage, a stem with eyes in it, hardening into a beak towards the front. This beak determines to a large extent the physiognomy of the particular genus and species. By its plumage we recognise the kind of bird, but the beak indicates its species.

The limbs, when featherless, appear wretched and stunted. Since the legs remain unfledged for the most part, they are often disappointingly gaunt. To the observer they seem poor, old, and helpless.

The arms with the feathers removed can only be compared to malformations, but when fledged they exemplify the power, beauty, and grace of the bird. Displayed in these wings is the true nature of the bird genus. They carry it through the air; they give it life in the element of warmth and light.

For this reason Gerlach is right when, in the introduction to his fine bird book, he writes:

> The birds are of lighter stuff than ours. They do not
> walk on heavy soles over the earth. Most of them touch
> the ground lightly with their toes, always ready to
> swing aloft. The air for them is space for unbounded
> movement, wherever they wish to go.[2]

This 'lighter' material is in fact a mineralised substance; for example, a bird passes excrement almost wholly encrusted with uric salts. Everything about the body is as if dehydrated. The skin has no sweat glands; the feathers are a mineralised callosity; the legs and toes are bony and sclerotic.

The body, right into the tubular bones, is filled with air through a system of air-sacs. Thus, the bird's body becomes a balloon which can easily raise itself into the air and float there.

For this reason its breath and the organs connected with it are the centre of a bird's life.

This body-build gives rise to the second activity – song – for which the bird genus is uniquely competent. No other creature comes near it on this capacity. Some species, such as parrots and budgerigars, ravens and starlings, can do more than sing: they imitate the sounds of the human voice so well that they achieve something like pronunciation. This has nothing to do with real speech: it springs from the bird's ability to relate intimately to the realm of sound. It can do this because sound is carried by the air, as is the bird itself. Sound and air, song and flight, are expressive of the airy world in which wind and warmth and light are in constant interplay.

In this element the birds have their existence. They surround the earth with their flight and song. To imagine the atmosphere without birds is an abstraction. The songs and cries, whistles and calls, croaking and screeching of the birds belong to the air, as do their flying and wing-beats, fluttering, hovering, and flitting. With all this the bird genus fills the air-space of our planet.

Feeding and nesting habits

The bird makes contact with the earth only in one place. Not where its toes just touch it lightly; that is a very fleeting and superficial contact. Hopping and tripping and putting one leg in front of the other are not true contact. These movements arise from the necessary search for food; not the bird but the beak itself makes these actions necessary. The nest, by contrast, is a structure which makes the bird an earth-dwelling creature. Nests are built from every conceivable kind of substance – soil and mud, twigs and small stones, moss and leaves, dung and sand. The form and type of nest is as varied as the species and genus of birds; each genus builds its own type of nest.

To this nest is entrusted the clutch (a number of eggs which is characteristic for each species) and then the time of brooding begins. Within the confines of the nest a cloak of warmth is created – a kind of oven in which the coming generation is sufficiently baked. It is an archetypal baking process that takes place here. What the elements themselves provide for in the case of fish and newts, reptiles and all invertebrates, is here attended to by the individual bird. The mammal has transferred this process into the female organism, where the womb has become an organic nest. The bird builds itself a structure which it turns into an incubator for a few weeks of the year. It then becomes the medium through which the bird makes contact with the earthly realm. Here the otherwise homeless becomes, for a while, settled and earthbound.

Two sorts of young emerge from the eggs. One group is unfledged and helpless. The other, covered with fluff and down, is perky and immediately ready for life and action. We refer to the first group as heterophagous (needing to be fed by others) and to the second group as autophagous (able to feed itself). Lorenz Oken, the great nineteenth-century naturalist, divided birds into these two great categories. But his ideas on this subject, developed at length in his massive work on natural history, are not always correct; between the heterophagous and the autophagous young there are many transitions, and his system of classification is no longer valid.[3] In this duality, nevertheless, there is hidden a valuable principle: we must only learn to understand it in the right way. This duality exists among many species right through all classes of animals, in the context of the first encounter of the newborn with the world.

Thus, for example, we have the embryo of the kangaroo, so dependent that it creeps into its mother's pouch and holds fast with its lips to her breast; or the beasts of prey, born blind and helpless, in contrast to the hoofed animals which are agile and quickly get on to their feet. A heterophagous creature is often a kind of larva which needs protection in order gradually to reach its final form.

In birds this process occurs through the development of plumage, which grows out over the small pitiable body, enveloping it by degrees until the bird is ready to fly. The feathers are of two types – soft down and stiff contour-feathers. The down feathers, more like a fluff which covers and warms, come first. The contour-feathers grow very gradually and form the actual flight-feathers. (The well-known ostrich feathers consist of down and so are of no use for flying.)

What then is it that separates autophagous from the heterophagous birds? Not simply the capacity, or the lack of it, to live without protection when they are born. It is something else, which points to the growth process of the individual species and is deeply involved with it.

The autophagous birds are much more closely connected to the earth than are the heterophagous. Autophagous birds do not need to wait until their plumage is complete before they can move. They slip out of the egg and at once begin their activities. The little chick pecks and hops about; it cheeps tirelessly, and although it follows the mother, it has already become independent. So it is that ducklings go into the pond and cygnets are so far advanced on the first day that they can be taken by the parents on to the water. Flying, if it develops at all, comes later and is then acquired. It is not absolutely necessary as the sole possibility of movement.

The heterophagous birds, on the other hand, are strangers to the earth. They are born into the nest as bundles of flesh. They would lie there helpless if the parents did not help to keep growth going. Their food has to be predigested and placed in their open beaks. The accumulating dung has to be removed by the mother with her beak and put out of the nest. On cool days the incubator is started up again and surrounds the young with warmth, since they are not yet capable of producing their own warmth. Only when the plumage, flight feathers and tail feathers are all full-grown is the heterophagous bird a complete bird, capable at last of living in its own element of light, air and warmth.

Hence we could speak of autophagous birds as being more suited to earthly life and more integrated into the earthly element. The heterophagous birds, by contrast, have remained more cosmic; they withdraw from the earth and want to connect themselves with it only when their plumage has fully developed.

The plumage is like a mantle drawn over a bird after it has slipped out of the egg. The feathers come as from outside, from the surroundings. For this reason Rudolf Steiner says clearly:

> The plumage is formed from without, and a feather can only come into being because the forces which work down upon the earth from cosmic space are stronger than the forces coming from the earth. The framework of the feather, what one may call its quill or spine, is of course subject to certain forces coming from the earth, but it is the cosmic forces which contribute what is attached to the quill and constitute the bird's plumage.[4]

These cosmic formative forces, which equip the growing bird with its coat of feathers, come from those regions from which the bird-beings have never quite separated themselves. For it cannot have happened that the birds in the past raised themselves from the earth into the air, developing from primitive reptile forms in such a way that in the course of countless years wings grew on them, and they then learnt to fly. Such a concept is not only unbiological; it is also – correctly viewed – quite untenable. The birds developed out of the air downwards towards the earth. They originated in the heights and have related themselves to a greater or lesser degree with the earth. An archaeopteryx was never an archetypal bird, but rather a bird-form which fell too far and thereby took on reptilian characteristics. It was the end of a developmental process, not the beginning. The archaeopteryx has died out because it could not live on. Unlike man and the mammals, the birds have never really taken to and accepted the earth.

Rudolf Steiner wished to indicate this when he said:

But in the bird nature we have beings that did not
assume the lowest functions; instead they overshot the
mark in the opposite direction. They failed to descend far
enough, as it were ... But as evolution continued, outer
conditions compelled them to solidify.[5]

In this way the fundamental principle of the bird world
becomes evident. They are creatures which have approached
earthly existence but have never been able to make it their own.
They are beings which have been involved in the process of
becoming animal, but have not carried this to its earthly conclusion. Had they done so, they would have thrown off their coat
of feathers, as do the seven ravens and the six swans in the fairy-
tales. So the birds keep their plumage, as do the mammals their
hair and the reptiles and fish their scales. They remain in their
enchantment, some raised too high, others sunk too low.

The autophagous birds were on their way to earth; they
sought it without ever reaching it. The heterophagous birds, by
contrast, were afraid of too strong a union with the watery and
earthy elements. So they have to lie helpless in an earthly nest,
until their plumage comes to their aid, raises them, and gives
them wings.

The swan

The great family of the swans, which inhabits many parts of the
world, belongs to the extensive order of geese and ducks. We
know some of the many species of this group; they range from
the Arctic over the northern regions of Asia, Europe and America,
and across the Equator into the southern continents. Only around
the South Pole are they absent. The eider-ducks, the divers, and
the mallards we know, and the domestic goose, to mention only a
few names. They indicate something of the rich variety of species,
colouring, and life-styles to be found in this order.

We are familiar with the activities of wild ducks on the ponds and lakes of the plains and in the mountains. They live also in the bays and among the dunes by the sea, and in autumn they fly in thousands on their journey south.

Their natural home is that boundary region between air and water where the air borders the reflecting surface of lakes, ponds, rivers and seas. All birds of the goose family can swim; almost all can fly, although flying is often harder for them than swimming. Especially taking to the air and landing are slow and laborious; the wings have to make enormous efforts, and the feet – sliding over the water – have to assist until a certain height is reached. Then progress is easy; with necks stretched out forwards and feet tucked in, the flight of ducks goes swiftly by.

A duck's beak is as long as its head, broadening out slightly to the front, and is covered with sensitive soft skin. The feet are short, often set far back, so that walking is akin to waddling. The four toes are webbed and give the feet a clumsy appearance. Walking is not the business of ducks; they are all swimmers and divers; their life is played out on the water. Only for the purpose of feeding, building nests and brooding do they go ashore; some even retire deep into the shelter of neighbouring woods or marshes or into hollow trees. The young are autophagous. A few hours after hatching they are ready to go into the water with the parents and they know exactly how to paddle and dive.

Swans are one part of this order. They are akin to ducks and geese, and yet different from them. They give an immediate impression of exclusiveness carried to the point of ostentation. One can understand that in Great Britain swans are the property of the Crown. They are royal visitors, envoys of a higher power. They have something unapproachable which surrounds them, as they glide over the water in their proud beauty. Then a swan's body, clothed in white or dark feathers, is like a boat which glides along, while head and beak are carried like a figurehead on the long, curved neck – a proud, often powerful sight, especially when the wings, slightly raised and extended like a shield, shelter the moving body.

Swans belong to the largest birds of their order. They are mostly monogamous; a mated pair may stay together for years. If migration occurs in the autumn (this happens only in the northerly regions of the earth), the swans will return together to the old nest, to keep the eggs warm for six to eight weeks, and to raise the young. Once the young swans have become somewhat independent, they are left by the parents and are treated as strangers. To become aware of their nobility, they must from now on master being alone.[6]

Swans are distributed particularly in the Arctic, the tundra and taiga areas of North America, in Europe and Asia, then there are the two species of the Southern Hemisphere in South America and Australia.

We know above all the humped swan, which lives on ponds and in estuaries of Europe. Large groups often take over a wide expanse of water. They seek the vicinity of human settlements. Small rivers which flow past old town walls, fish-ponds near monasteries, quiet ponds in old gardens, remote estuaries on which old villages nestle – these are their home. The snow-white plumage, the red beak, and the dark hump on the beak give this swan his aristocratic bearing.

Further to the north live the somewhat smaller whistling swans *(Cygnus musicus)*. They belong to the species of which the Greeks used to say that they came with Apollo to Delphi every year from the north; the god was swept along on their wings, bringing tidings from the land of the Hyperboreans. Many observers tell of the strange, sometimes bell-like voice of this species; and it is said of them that they sing when their comrades are about to die (the origin of the 'swan song'). During the winter they leave their home in Iceland and Scandinavia for the regions of Central Europe. They are also at home in Russia and Siberia.

The black-necked swan nests in South America, from Peru to the Falkland Islands, and in Brazil. Its plumage is white, but the neck and head are covered with black feathers. The hump on the greyish-yellow beak is red, and the wings are short, yet it is a good flyer.

The dark-coloured mourning swan lives in southern Australia and Tasmania. Underneath a covering of blackish-brown feathers the white flight-feathers stand out, and speak of its brothers living in the north.

This picture of their geographical distribution shows clearly that swans belong to the north. The whistling swan lives right up in the Arctic region. The temperate zones are the habitat of the humped swans. The tropics are uninhabited by swans, and further south, we find only the two dark-coloured species of South America and Australia: the black-necked and the mourning swans. This distribution gives an impressive picture of swans all round the earth.

Although swans are water-birds, they need the land for nesting and breeding. But they very seldom take off from land, and only by running for a long distance with outspread wings; they use the water as their runway, and they always land on water, though they have been seen to take to the air and to land on ice.

It is understandable that these noble birds have been held sacred. The beauty and dignity which they radiate call forth in us a feeling which points to something higher. The swan is by no means a gentle lord. He is quick to attack, strong and easily aroused to anger. He then lets fly in a rage with wings and beak, and even takes on more powerful opponents. His white plumage gives him an aura of invincibility; his courage makes him a knight.

Do we not feel, when we look at the swan, that there are higher regions of soul than the everyday? In ordinary life, we are geese and ducks, stupid and unconcerned, joyful or painfully moved. But beyond all this lives the swan in us. The exalted bird of the soul which comes from the far north and often visits us only as a guest before going on further. In the Middle Ages, if we wished to dedicate ourselves to this idea of the swan, we would have become a member of one of the many orders of Swan

Knights which existed then. The bird on their coat of arms was the white swan, which enjoined the knights to raise themselves above the goose and duck existence, and to lead their lives in proud service.

The stork

A quite different air plays round the stork. What a difference between the swan gliding silently along and the stork stepping across marshy meadows! The swan hides its short clumsy legs, which are useful only for paddling. The stork walks as if on stilts, which are painted red to make them stand out. The stork's beak is as long as its legs, and often so big and heavy (for example in the marabou) that it tilts the bird's head downwards.

A general feature of the build of birds is indicated here: the shape and size of the beak and of the legs correspond to each other. If the beak is small, the feet and legs are also small. The hard curved beak of the birds of prey is reflected in the claws. This harmony between two distinctive features of birds is especially evident in the stork.

It is a strange order to which the storks belong. They are referred to as Ciconiiformes or Gressores, a name which points to their special feature – their gait. Four families make up this order: herons, ibis, hammerheads, and storks. These birds are all essentially of one type, with relatively long, thin legs, and long, pointed beaks. Neck, head, and beak are almost like a limb which seizes food with fast serpentine movements. The Gressores are birds of prey. They eat everything which moves around them: frogs and small toads, worms, lizards, beetles, mussels, fish, even young birds, young hares, moles and mice. A blow or a stab with the beak and the prey is quickly consumed. This behaviour belies their appearance; storks look tamer than they really are. But the red colouring which occurs in various parts of the body in some species gives a hint of their aggressive, warlike nature. One could classify them as cruel melancholics. Shallow ponds, lakes or rivers are their hunting-ground. They live on the shores or river banks among reeds and rushes, papyrus and willows.

These birds, with few exceptions, build their nests high in the tops of trees or on roofs, as does the stork which has become a close neighbour of man. Large branches and twigs are collected for the nest, which is usually round, and lined with moss and dung, straw and leaves.

The young are heterophagous. During the first weeks they lie there helplessly; they have to be fed and cleaned by the parents and also need the parents' protection against excessive cold or heat. We owe to Horst Siewert such detailed observations of black and white storks that we are familiar with this process in every detail. He describes how frogs, which have been predigested in the stomach of the parents, are regurgitated, cut into small pieces and pushed into the open beaks of the hungry young. We are told also how on cool evenings, the parents sit on the nest so as to warm the young; on hot days, when the sun shines on the eyrie, they perch on the edge of the nest in such a way that the nestlings lie in their shadow.

It takes several weeks for the contour-feathers to grow and for the legs to be strong enough to take the first steps on the edge of the nest. The young black stork, with its clumsy wings, then flies quickly from one branch to the next in the tree where the nest is, and returns to the nest. Until one day, suddenly, parents and children set off together to hunt for prey.

The Gressores, including the heron and the ibis, are at home all over the world, except for the far north. Their habitat is always a region between shallow water and land. Where marshes and fens are formed and the earth element and the water element meet, there tread the feet of the Gressores, and their beaks penetrate into this area of organic life. They have, as it were, descended one level lower than the ducks, for ducks come down only as far as the boundary between air and water.

However, it is only with foot and beak that the Gressores

penetrate where earth and water meet; otherwise they remain connected always with the air. Then for building their nest they rise to the tops of the trees, there to wait and care for their young.

The storks are the only birds in this order which at least attempt to draw near to the earth by coming close to man, as the white stork does. They build their nests on the roofs of houses and stables and return year after year to the old nest. Children love them and except them to bring them brothers and sisters. And the adults smile at this superstition or condemn it as nonsense.

Can we ascribe to this presumption of the intellect the fact that the stork is gradually ceasing to frequent the dwellings of Central and Northern Europe and withdraws ever more from them? We know very well that the stork does not bring our children, but what was it that gave rise to this inner picture for such a long time? Siewert describes an experience which might perhaps put us on the track of an answer:

> In a small Pomeranian town the storks appeared just as school had finished. The children poured out on to the streets, and a little lad who soon spotted the large birds cried out his discovery to the world. All eyes were raised, and all the children laughed at the sight of the long-legged birds in the clear bright sky. But it was not only the children who were glad; many people gazed up from the narrow streets and forgot that only the day before yesterday the last snow had fallen and that the air was still bitterly cold. Even if the people did not sink to their knees at the first sight of the storks, as they had done two thousand years ago in honour of the bearers of the spring, at least the joy has remained to this present day, for, just as in ancient times, the wanderers brought the spring to this northerly land, with the sun and warmth of the south, and the long winter with its terrors was forgotten.[7]

When Siewert says that the storks were the bearers of spring he is quite right, but two or three thousand years ago people fell to their knees at the sight of the birds because they knew that with them the souls of unborn children were approaching the earth, and that the time of pairing was beginning. Out of his spiritual insight Rudolf Steiner indicated that up to the first pre-Christian millennium births among the Germanic tribes were so arranged that they mostly occurred around the time of Christmas.

This mythical truth was spurned and spoiled by the intellect during the nineteenth century, until it gave rise to the stupid and indeed ridiculous pictures of the stork carrying the newborn baby in its beak. But behind this we can discern the true announcement which the stork as messenger of the spring once brought to the people of the north. His clattering sound is heard at the time of weddings, and when on Good Friday children parade through village streets with rattles, they remind us of times past.

The white stork was – and sometimes still is – accompanied by his darker brother. The black stork, however, keeps away from people. It builds its nest in certain tall trees in the depths of forests. In the autumn it returns, like its white brother, to Africa. The migrations of the storks have now been well researched. Avoiding the Mediterranean, they pass from north to south on two main flyways. The eastern route leads across Bessarabia, along the Black Sea to Asia Minor, then over Syria and northern Arabia, across the Red Sea to the Sudan, and from there through eastern Africa to various parts of southern Africa. The western route passes over southern France and Spain to Morocco and Algiers and across the Sahara mainly to Senegal and the Niger.

No other storks go on this long journey, which covers a good part of the globe. Those living in the Sudan and on the Blue and the White Nile do not migrate. Storks such as the marabou also remain where they are and roam around only at specific times of the year.[8] The storks of this family, with unfledged throat and gigantic craw, feed primarily on carrion. Like the vultures (which also have unfledged throats) they seek for game which has been killed and left by other carnivores. All the storks which live in Africa and India have strange forms. Compared to them, the white stork is like a child which has not yet been stricken by the darkness and need of the earth, but has kept its original purity. It is nearer to the archetypal form of the group-being of the storks. The other storks have become too entangled in the muddy realms of the earth. The marabou storks seem to have sunk the lowest. That is why they eat carrion and could be called the hyenas among storks.

The more domesticated stork stands apart. It leaves the regions of Africa, which are well provided with food, and journeys every year to the north, to bring its message to the people.

Since humanity has started to tread the path towards freedom, however, and their children are born at all times of the year, the mission of the stork has come to an end. Since the beginning of the twentieth century they have been disappearing from Central and northern Europe. They have now begun to settle in southern Africa, and many individual families of storks have been found nesting there. Will they make this land their habitat for ever and forget Europe? The destiny of the group-souls of animals and birds is as diverse as that of individual men. Only sometimes is it granted to us to get a glimpse of their real task.

The message of the swans and storks

Among the fairy-tales collected by the brothers Grimm is the tale of the six swans. It tells of a king who lost his way while hunting deep in the forest and came to a cottage where a beautiful girl was waiting for him. But her mother was a witch, who forced the king to take her daughter as his bride; only then would he be shown the way out of the dark forest. And so it happened; the king took the beautiful daughter of the witch as his wife, but hid the children from his first marriage – six boys and one girl – in a lonely castle. He was shown the way to the castle by a magic thread, with the help of which he was often able to visit his children. By treachery the wicked woman gained power over the magic thread and enchanted the six boys, who encountered her on the road to the castle, turning them into six swans. The girl, who had stayed at home, was saved from this transformation.

That is the first part of the tale, itself formed from two stories interwoven with each other. The first tale tells of the king, of the man who becomes ensnared in the earthly world of the dark forest and is forced to unite himself with the beautiful appearance, the daughter of original sin. The holy offspring from the first marriage, which was free from sin, are kept in an inaccessible castle. Is not this castle, to be reached only by means of a magic thread, the far north, the mythical land of the Hyperboreans? Magic can indeed enter there and transform the children into swans, but it cannot destroy them. The six enchanted brothers are hidden powers of the soul which are to be redeemed one day. On these powers of the soul – the swans – Apollo travels every year from the north to Delphi, to bring powers of the sun and of renewal to men.

They are the same powers which were acquired by initiates of the third degree, who were called Swans for this reason. Rudolf Steiner says:

> The third stage of initiation is that of the swan. A 'swan'
> is he who has progressed so far that all things speak
> to him, even those which have their consciousness on
> higher levels ... One must rise to higher worlds to find
> the 'I', the names of the other beings. These things speak
> their own names ... The swan initiates were no longer
> allowed to bear their own names, but the whole world
> revealed its names to them.[9]

This helps us to understand the second story in the fairy-tale. The sister of the six swans is given the task of remaining silent for six years – a year of silence for each brother. At the same time she has to sew six shirts from star-shaped flowers. The girl decides to perform these tasks. She goes into the forest, finds herself a seat in a high tree and begins the work. A second king comes with his hunters into the forest; they find the girl and, although she resists, she is fetched down from the tree; the king takes her up on to his horse and makes her his wife. The girl maintains her silence, although the king's wicked mother constantly abuses her and takes away her newly born children. She remains true and loyal to her swans. At last, as she is standing at the stake to be burned, the six years are over. The six swans fly down; the six shirts are thrown to them, and the six brothers are once again united with their sister. Now she is able to speak and to convince her spouse of her innocence. At this, the destiny of the swans is fulfilled.

The obligation to remain silent is found also in the Lohengrin saga. The saga tells of the son of Parsifal who is led by a swan into the land of Brabant in order to establish peace there. He is also under a rule of silence, though he has to be silent only about his name.

The tale of the six swans tells two stories. One deals with the enchantment and the other with the redemption of six brothers. The enchantment occurs through the wife of the first king; the redemption through the spouse of the second king, the sister of the six swans. The first part of the tale is a story of pre-Christian man; there the saga of Apollo appears in disguise. The second part reaches into the Christian era and tells the Lohengrin saga in the form of a fairy-tale. In both parts the swan is mysteriously involved.

There is an indication by Rudolf Steiner in which the deeper background of the medieval swan knight orders is explained. Here we are told how, during the first few Christian centuries, the barrier between the living and the dead was not so impenetrable as it later became:

> The dead remained among the living. Outstanding,
> revered personalities, during the first period after their

death were undergoing, as it were, their novitiate in sainthood.

For the people of that time it was not at all strange to speak of the living dead as of real persons. A certain number of the living dead, especially elect ones, having been born for the spiritual world – were appointed as 'Guardians of the Holy Grail.'

Then it is explained that some men here on earth became representatives of these guardians, and they were united in the various Orders of the Swan. 'They were persons who wanted the Knights of the Grail to be able to work through them in the physical world.'[10]

That is how a picture of the swan presents itself in connection with the work of the illustrious dead. Its essential nature refers to something departed, something lost to man on earth. The soul feels that it has become a duck or goose in the earthly body; that the swan nature has eluded it, and that the seeking and recovery of this has become its task. The soul identifies itself with the sister of the six swans and can find in her example the power to survive in the pain and poverty of the earth until the lost swan brothers appear to her. So it was that Lohengrin turned to Elsa of Brabant and served in the forces of the Duke of Brabant, Henry I, in the battle against the Hungarians. Since then many a swan knight has appeared in times of need and affliction as a messenger of deliverance. The soul can then sew further the shirt of 'star-flowers', so as to be ready for the heavenly wedding at the end of the time of probation.

In pre-Christian times the existing powers of the soul were enchanted. They became the six swans, which in terms of the old initiation could become disenchanted only at the third stage. In Christian times, however, some succeeded in becoming the bearers or vessels of the exalted dead, and this sometimes without inner schooling. By this means they became messengers, Knights of the Order of the Swan.

In the medieval Lohengrin epic there is a passage which brings out clearly this Christian power of the swan. As the swan is drawing his lord over the sea in a boat, Lohengrin asks him for food. The swan dips his head beneath the waves:

> As if he perceived fish.
> Look there how a small shoal
> Is borne by the swell into his mouth.
> The knight saw them as dry and clean,
> The swan passes them to the hero with his beak.
> He takes them with eager joy,
> And eats one half and gives the swan the other.
> Never were bird or lord so well fed.

After this repast, the swan begins to sing and now Lohengrin can recognise: 'This is in truth an angel pure, who floats with me upon these waves.'

The stork presents us with a different picture. Its venerable brother, the ibis, was sacred for the ancient Egyptians. They respected it so much that they would embalm the bodies of fallen ibises and lay them in special graves. The death penalty applied to the killing of an ibis, even if this was unintentional. The thrice-great Thoth, whom the Greeks called Hermes Trismegistos, was often represented with the head of an ibis. Even his hieroglyph was a stylised ibis. On his head the Thoth-ibis wore the crescent of the moon, in which the disc of the sun was poised. Thoth was the initiator of Egyptian culture. He was the god of speech and writing and is often represented with a stylus in his hand.

What the ibis was for the Egyptians, the stork became for the peoples of the north. He was wise enough to indicate to them the coming of unborn souls, who were ready to be incarnated on earth. Although, like the swan, he wore a white garment, he evoked a quite different image for the soul. His were not powers of the heart but powers of wisdom. To the soul he appeared not perpetually young, like the swan, but old and clever like a midwife.

The swan is connected with that which is gone, the stork with the unborn, with that which has not yet come. Ibis and Thoth were related to the moon; the swan to the realm of the sun. So it was that Apollo's swans came out of the realm of the Hyperboreans in the north.

The storks came every year from the southern moon-regions of the Lemurian zone up to the north of Europe. Wisdom and humility were balanced in these migrations. The ascending souls of the dead, while on the way to the realm of the sun, were in the region permeated by the spirit of the swans. The souls returning to earth from the sphere of the moon were related to the storks.

The swan's capacity for sacrifice allowed it to affirm its relation to the earth; it became an autophagous bird. The cleverness of the storks kept them back from too strong a connection with the earth. They remained heterophagous.

That is how these two species come before us, both in their contrast and in their joint effect on the soul. They are like memory-pictures which have remained alive from a past period of human history. Once they were united with the souls of men. But the earth-substances became hard and impenetrable, and only a few human souls could take up and fill out their bodies. Then came the birds: hard horn grew into their etheric wings, which became earthly organs and made it possible for them to encircle the planet. Nevertheless, they remained united with the sun, and thereby with the human soul-realm.

From then on their destiny took manifold shapes and brought them many tasks. Many were lost; others became songsters; many attached themselves to men, as did the chickens, the doves and pigeons. The storks and swans remained connected to the higher part of the human soul. They indicate human destiny: to be born and die, to be wise but to bear humility in the heart, and one day to become – perhaps – a swan knight. Then the stork will be redeemed.

7. The Dove as a Sacred Bird

The dove in the Bible and in history

Wherever in human settlements there is an open space, in squares and streets, in yards and gardens, we find pigeons. They coo and fly, hover and patter before San Marco in Venice just as they do round Nelson's Column in Trafalgar Square, in the Tuileries in Paris, or in the Rathausplatz in Vienna. Many a worthy farmhouse had its dovecot, and we find the same at the centre of many a little town or village, not only in Europe, but in Asia, Africa, and America as well – wherever human beings gather in permanent settlements. There are many kinds of domestic pigeon, and they appear to have been the companions of humanity for many thousands of years. Their existence is deeply rooted in the conceptions of mythology, and can be followed far back into the first beginnings of human history.

In the Old Testament the dove appears in connection with the end of the Flood. Noah, when the first mountain tops had appeared above the sinking waters, and the raven had flown to and fro, released a dove. But she returned, for she 'found no place to set her foot'. Seven days later, when another dove had been sent, she 'came back to him in the evening, and lo, in her mouth a freshly plucked olive leaf.' But Noah still waited, and after seven days let another dove fly out, which did not return to the ark; 'and Noah removed the covering of the ark, and looked, and behold, the face of the ground was dry.' (Gen.8:9–13).

Later the dove appears again in the Song of Solomon (2:14): 'O my dove, in the clefts of the rock, in the covert of the cliff, let me see your face, let me hear your voice, for your voice is sweet, and your face is comely.' Here she appears as the intimate symbol of the World-Soul, uniting with the awakening human 'I' as comprehension through knowledge.

In the Psalms (55:6; 68:13) it is said: 'O, that I had wings like a dove! I would fly away, and be at rest.' Or 'Though they stay among the sheepfolds [they gleam like] the wings of a dove covered with silver, its pinions with green gold.' Here is the lifting up of the soul, which leads out of the trouble and distress of this world.

In the prophets the dove appears once more. We read in Isaiah (38:14): 'Like a swallow or a crane I clamour, I moan like a dove. My eyes are weary with looking upward. O Lord, I am oppressed; be thou my security!' And Jeremiah (48:28) says: 'Leave the cities, and dwell in the rock, O inhabitants of Moab! Be like the dove that nests in the sides of the mouth of a gorge.' Here the dove appears as a fugitive, returning from the towns to the rocks of the coasts, from which it had once come to the human settlements.

In the religions of the Near East the dove was held sacred. The goddess Ishtar in Babylon was specially connected with it, and doves were sacrificed to her. Doves were consecrated in the same way to Astarte. They drew the chariot of Aphrodite; Venus Anadyomene was hatched from an egg by a dove. At the sanctuary of Zeus at Dodona, in northern Greece, doves lived in the sacred oaks and gave the answer, as holy oracles to the questions of the faithful. In the Phoenician language the same word was used for dove and for priest; in Hebrew the word for dove is the same as the Arabic word for priest. When Herodotus says that Phoenicians once brought a priestess from Thebes to Dodona, it could be either a priestess or a dove that is meant; the words are identical. It is known too that in the festivals of Adonis doves were burned in honour of the god; and Aeneas was led by doves to the Golden Bough.

From all these indications it can be seen that doves accompanied men in their migrations, that they were held sacred in connection with sacrificial ritual, and that they were attached not only to human habitations but also to temples and places of the mysteries.

Habitat of doves and pigeons

All domestic pigeons are descendants of one species, the rock-dove, *Columba livia.* This species is distinguished by its way of life. Wild pigeons in general belong to the woods, living at least part of the time in trees; they are to be found all over the earth where there are woods or regions rich in trees. Only a few inhabit cliffs and rocky terrain. Neither in the far north, in tundra or steppe, nor in mountains beyond the tree limit are there pigeons. Everywhere else they are at home. More than three hundred species are known.

Only the rock-dove does not need woods. It lives on cliffs and rocks, and in ruined buildings, as the prophet says. It is found in Europe on some northern islands. It is especially numerous in County Donegal in Ireland, and lives all along the West Coast of Scotland, the Hebrides, the Orkneys and Shetlands, the Faroes and the small rocky island of Rennesøy near Stavanger in Norway. It dwells in rocks and cliffs round the Mediterranean; in Greece, Spain and Italy, France and North Africa. It is at home in the Levant and Syria, and lives all over Asia Minor and Persia, reaching as far as the Himalayan region. If one considers the distribution of the rock-dove, it looks as if this bird occupies the great paths of human migration. From India through Persia to Asia Minor, and then along the Mediterranean coasts, everywhere rock-doves are to be found. Through thousands of years the post-Atlantean civilisations moved in this way from east to west. The rock-doves found in these regions, however, are often domestic pigeons which withdrew from human settlements and returned to a free life. Brehm writes:

> In Egypt I saw them on cliffs, particularly near rapids.
> In India they are among the commonest birds, laying
> their eggs in holes and on ledges in rocks and cliffs,
> if possible near water. Here, as in Egypt, they are in a
> half-wild condition and occupy old, quiet buildings, city
> walls, pagodas, rock temples and such edifices.[1]

From this description it can be seen that between the rock-dove and the domestic pigeon there is a gradual transition, and that the domestic birds can easily pass over into their wild or semi-wild condition. Certainly the main division among pigeons is not between those that are wild and those that are domesticated, but between those which live in woods, and those which have their homes in rocks and in human settlements.

Rudolf Steiner often indicated that the peoples who migrated eastwards after the decline of Atlantis crossed Europe and Asia by two great routes. Through this there later developed two distinct, widespread peoples; the Iranians to the south, the Turanians more to the north.

> Thus there arose what is perhaps one of the greatest
> antitheses in the whole of post-Atlantean evolution: the
> antithesis between these more northerly peoples and
> the Iranians. Among the Iranians the longing arose to
> take a hand in what was going on around them, to live
> settled lives, to acquire possessions through effort, in
> other words, to apply man's spiritual forces in order
> to achieve the transformation of Nature. That was the
> strongest urge in the Iranians. And in the immediately
> adjacent lands to the north, lived the people who saw
> into the spiritual world, were on familiar terms, so to
> speak, with the spiritual beings, but were wanderers,

having no inclination for work and without any interest
in furthering culture in the physical world.[2]

Here there is a similar polarity to that described among the
pigeons. The wild wood-pigeons are like the Turanians, and
the rock-doves, which can become domestic pigeons, are like
the Iranians. If we think of Noah's action in sending out the
three doves at the end of the Flood, we see in an image the
beginning of the migrations to the east; the differentiation
into the northern and southern groups is present also in the
separation of the two main groups of pigeons. It can indeed be
assumed that the wood-pigeons accompanied the Turanians,
the rock-doves the Iranians, on their journey from west to east.
Thus the rock-dove was tamed and domesticated, and made its
home both in the sacred and the profane parts of the Iranian settlements when these became established. Doves have thus a close
link with human existence.

Feeding and flight

The fact that the Iranians became settled led to the emergence in
the domestic pigeons which came to live with them, of a peculiar
characteristic; the capacity to find their home abode again from
hundreds of miles away. Many other birds have this capacity,
but only when they are carrying out their long annual migrations. The pigeon can do it, with a little training, at any time of
year. Thus carrier pigeons were used from antiquity; they were
already known in Egypt. When Rameses III came to the throne,
the news was sent all over Egypt by carrier pigeon. Besieged
cities used carrier pigeons to communicate with the rest of the
world, a method rendered out of date only by wireless telegraphy
at the beginning of the twentieth century. As recently as the siege
of Paris in the war of 1870–71, many reports were brought to and
from Paris by carrier pigeon.

This transmission of messages through a kind of homing instinct, which is so deeply rooted among pigeons that they can find the way back over great distances, even at night and through regions entirely unknown to them, is a very remarkable characteristic. The peacefulness, gentleness, and the family sense of pigeons so often noticed, are related to this. It is hard to say, as the observations are not definite enough, whether it is really true that a pair of pigeons remains together inseparably over the years. [That is considered to be the case now.]

But one thing is certain; among all the birds, the various kinds of pigeon have a characteristic belonging to them alone: They feed their young with a milky juice, like mammals and human beings who feed their newborn with their own substance, milk. Among the pigeons it is not only the mother which forms this substance, but both parents. No other bird has this characteristic. Many indeed have the organ, the crop, in which dove-milk is developed – but only pigeons produced it.[3] The crop is an often considerable widening of the alimentary canal, at about the point where the neck meets the breast. In other birds, and pigeons as well, it is used to mix food with saliva, and to begin the digestive process.

With pigeons, however, in the middle of the brood period the crop begins to swell, and to increase in size. Gradually the inner layers of the mucous membrane of the crop become so fat, that they desquamate, and the mass of cells is gradually dissolved in the cavity of the crop and becomes a kind of white broth, which is given to the young brood as food. This 'milk' is rich in fats and proteins, and for about three weeks it is the only food of the young pigeons; after that can they begin to take grain.

It could easily be said that it is not a real milk, but only something popularly described as such. But there is a very significant fact on the other hand. Mammal and human milk, which is formed in the mammary glands, is connected with a hormone produced in the pituitary gland. This hormone, called prolactin, stimulates milk production considerably; the supply of prolactin equips the milk glands for the production of milk. But the same substance can stimulate the pigeon's crop to form dove-milk and this happens so regularly that this method is a quantitative test for prolactin.

Thus there can be no doubt that dove-milk is related to mammalian milk not only by its name, but belongs to the same group of substances and is a real 'milk'. It can be assumed that the close family relationship among the pigeons, their readiness to settle and their peacefulness are connected with this formation of milk. For the young pigeon receives a food-stuff peculiar to its kind, which attaches it much more strongly to the protection of the nest and to the family than is ever the case with other birds. A part of the bodily substance of the parents is given to the young brood over many weeks, and the blood-relationship with the whole species is thus deeply influenced. The milk is a living bond holding together through the generations of doves as a species in the most intimate way. Thus the rock-dove can change so easily into the domestic pigeon, and back again.

The crop and the larynx

In antiquity doves were called the 'guests of the gods'. They chose as their abodes temple buildings above all, and settled in the holy precincts of the mysteries. Wherever in Asia, in Europe and in North Africa temples stood, they were continually encircled by the flight of doves. The rock-dove with its brilliant grey-blue and green feathers became gradually the white temple-dove, which was used as a sacrificial animal in the most varied civilisations. In the New Testament too we find references to this. Thus doves became not only companions of people, but were drawn into the holiest rites that can be performed. In the groves of mysteries the doves settle, and make their homes around the holy buildings.

They were messengers, and brought over great distances what human beings wished to communicate; this was only expected of them because they were felt not only as guests but also as messengers of the gods. And as they brought tidings from the gods to human beings, people used them as messengers among themselves as well.

Rudolf Steiner described the task of birds in general in the cosmos, showing that it is their vocation to spiritualise matter:

> One can actually say that, when the earth has reached the end of its existence, this earth-matter will have been spiritualised, and that the bird-creation will have had its place in the whole economy of earthly existence for the purpose of carrying back this spiritualised earth-matter into spirit-land.[4]

If this is the work of the birds in general, then a particular species must surely have the task of transforming special kinds of earthly substance back into the spiritual. Could it not be that doves have a particular duty? They give their company to human settlements, and are thus concerned with buildings and houses. They are connected with substances used by humans. For wood, stone, mortar and everything else used in building is altered by the work spent upon it. And all these buildings are permeated by the human *Word*. In and around the houses and temples, the grave-monuments and the palaces, human speech is used, and unites with the walls. And doves become the messengers who release the human word, in its endless variety, from its enclosure in matter, and embody it again in the spiritual substance of the cosmos.

All that has been spoken, all the good and the bad that is clothed in words, is reunited with the cosmos through the function of the doves. Their work as messengers is only an earthly picture for their cosmic task. They are bearers of the Word, and they are prepared for this in their infancy. The milk given by father and mother holds them to the work that is to be done. For the bird's crop is situated anatomically where the larynx is in human beings. In the larynx the word is produced, but in the pigeon's crop milk.

Mammalian milk has a definite task; it enables the newborn child to form bone substance in the right way. For milk is not just a general means of nutrition; it is specially prepared to form the material for the skeleton. Through receiving milk the human child becomes a citizen of earth; the mineral scaffolding of his bones is hardened by milk, and becomes the rock sustaining his existence.

The milk produced by doves has the task of anchoring their mission in the realm of the human Word. The skeleton is intimately connected with speech; only the human race, with a skeleton which has become the image of the whole universe, can speak. The human head is round like the universe above; the ribs follow in their form the paths of sun and planets; the limbs are like pillars, permeated with the forces of the earth. This perfection of shape makes possible a larynx which can become the cradle of the sounding Word.

Human milk, arising in the mammary glands, provides the material substance from which the bones, an image of the whole universe, can be shaped. It is a material which conforms to the cosmic forces, and serves them. This enables the sounding Word to speak in the human being.

Dove's milk, arising in the realm of the larynx occupied by the bird's crop, helps towards the redemption of the spoken Word, which has united with the matter shaped by human hands. From the grave of matter the dove liberates the human Word.

Human being and dove become companions who have journeyed together into the land of earth, to become servants of the Word. Therefore doves are the 'guests of the gods'. For this reason the words for 'dove' and 'priest' are almost identical in many ancient languages. For this reason Noah sent the dove out of the ark, to find out whether the earth had become as hard as bone again, so that it could be trodden by human feet. For this reason the dove is 'the darling of her mother, flawless to her that bore her' as the Song of Solomon declares (6:9).

The dove and the Word

The dove appears too above the head of Jesus of Nazareth, when John baptises him in the Jordan. Rudolf Steiner says about this:

> While in a physical incarnation something spiritual descends from higher worlds and unites with the physical, that which was sacrificed in order that the Christ Spirit might enter appeared above the head of Jesus of Nazareth in the form of a white dove. Something spiritual appears as it detaches itself from the physical. That is an actual clairvoyant observation and it would be far from right to consider it a mere allegory or symbol. It is a real, clairvoyant, spiritual fact, actually present on the astral plane to clairvoyant sight. Just as a physical birth implies the attraction of spirit, so this birth was a sacrifice, a renunciation; and thereby the opportunity was provided for the Spirit, Who at the beginning of our earth evolution 'moved upon the face of the waters,' to unite with the threefold sheath of Jesus of Nazareth and to strengthen and inspire it through and through, as described.[5]

The dove appears here as the result of the sacrifice through which the Logos, the Spirit, who brooded over the waters (Gen.1:2) could draw into a human body. Thus here too the dove becomes a helper of the 'Word'. The Logos permeates the body right down into the bony system, into the earthly substance which is prepared by milk. At this holy place of human history, at the Baptism in Jordan, the true being of the group-soul of the doves is revealed. Through its sacrifice the place is prepared which is to be filled by the cosmic Word itself. Until then it has been guest and messenger of the gods; now it gives up this task, for the Logos itself has assumed it. Perhaps once, when people built the Tower of Babel, and their original language suffered the fate of fragmentation, doves began to prepare to lead the divided, splintered human word back to the realm from which it once came. But then the Logos itself entered the earthly world, and the task of the doves was thereby completed for the time; their priestly duty carried out, their mission fulfilled.

Today they have become commonplace. The mystery temples are ruined, and so pigeons live in squares and marketplaces, a picture of how the word is wasted and misused. As in the past, they pick up words, keep them and bring them back into the spiritual world. They have become messengers of men. In German they are called *Taube*. Etymology does not connect this, as might seem likely with *taub*, deaf, dull, 'dumb', but regards it as imitative, representing the coo of doves.* However this may be, the pigeon today seems a stupid creature, because its true being and character cannot be recognised

In the future its light will shine again, and reveal its glory where sacrifice and ministering messenger-service count for more than powerful appearance. The words, 'I was naked, and you clothed me' (Matt.25:36) applies to the doves. They gave back their function, their work in the service of the Word, to him who as Logos came unto his own.

---

* Eric Partridge, *Origins*, says that 'dove' and 'deaf' are perhaps connected.

8. The Sparrows of the Earth

The life of the sparrow

Wherever people live and have settled the sparrow is at home. In the many and various names used for him in Germany* we can hear the sympathy but also the slight disrespect with which people regard him. Few are really friendly towards him. One puts up with him as a habit; who bothers about sparrows? They are so numerous and commonplace that they are hardly noticed. And yet they are our companions and settle wherever humans are established.

There is hardly a farmhouse, village, market-place or suburban street which has no sparrows. And the more densely people live together, the more noisily do the sparrow hordes bustle about. In the midst of big cities, on main roads, in backyards and gardens and in the smaller parks they are at home. Where people crowd together, where children play, vehicles clatter and lovers embrace, where life and death daily touch hands – there also the sparrows twitter.

Grey, almost like ashes, is the feather garment of the house-sparrow. Around the eye and beak it is somewhat darker, with a brownish stripe running from the back of the eyes – at both sides of the head – towards the neck. The wings are brown and dark on top, blackish brown on the underside. No yellow or red, no blue – only a few white patches adorn the little bird. He is really insignificant and without special markings.

But when you know him more intimately and make friends with his appearance, his habits and peculiarities, his deeds and

* *Hausspatz, Haussperling, Hofspatz, Rauchspatz, Dieb, Sperk, Hausfink, Mistfink.*

doings, then he becomes lovable and quite remarkably friendly and interesting, for he knows life and fits himself as modestly into it as circumstances allow. He possesses his nest for himself alone; no-one may cross his path there. But he does not object if another sparrow builds a nest near him and moves into it. Husband and wife – if they live long enough – often remain true to each other for years for they stay by the nest they have set up or bound for themselves. The nest is home, security and possession.

As much as they are preoccupied with their own nests, just as much do they like going about in the company of other sparrows. One will seldom find a single sparrow. Wherever they loiter around, they appear by the dozen. They are so similar that it is difficult to distinguish them from one another. Their games, movements and reactions are almost identical. They like to quarrel, chatter with one another, hop, tumble about in the sand and puddles of the street, whirr over the pavement and suddenly fly off, back to the nest, soon returning to rejoin their brothers and sisters.

Just as ordinary as their colours are, just as simple and limited is the rest of their bird-existence. They can chirp and twitter a little, but it is not given to them to sing. House-sparrows also make no great excursions. In autumn they stay in their nests; the unrest and fever of migration is alien to them. At the beginning of August, or somewhat later, they go away, always together with their wives, for a little holiday into the nearby countryside, mostly near ripening cornfields. They rent a summer residence in the hedges surrounding the fields and there they play and peck, bathe and sleep. On some days the pair go back to the nest to see if everything at home is still in order; but soon they appear again in the fields.

If these summer guests are very numerous, the farmer will have difficulty in bringing in a full harvest, for sparrows can eat without stopping. But if there is not much grain, all kinds of beetles, grasshoppers, caterpillars and worms are also consumed. But only the quite young sparrows like such animal food; the adults prefer corn and other seeds.

When autumn comes they all fly back to town. The waiting nest is newly upholstered and lined and prepared for the winter. Seldom do coupling and egg-laying come about at this time. Only in really warm autumn weeks is there still some springtime romance. It is more like a second summer, mostly without visible results.

Sparrows can have quite long lives if nothing cuts them short. Their average life-span is seven to eight years. Some reach their eleventh or twelfth year before they die. But the nests last longer. If one member of a pair dies, the nest will provide for a male or female joining the survivor. And so it goes on, a kind of life-quadrille, with the nest always as its centre.[1]

Nesting and habitat

In Gaston Bachelard we read:

> When we observe a nest we find ourselves at the source of a trust towards the world, we touch on a focal point of trust, we are struck by an appeal to cosmic trust. Would a bird build its nest if it did not have an instinctive trust in the world? ... In order to experience this trust ... we do not need to count up the material reasons for it.[2]

For they hardly exist. What is at hand is based – with people as well as with a bird – on this 'cosmic trust', which ever anew goes about setting up a nest, a house, a dwelling, a home of one's own, where protection is sought and also found.

The place of security, the feeling of being protected, is one of the original human needs on earth. Many birds also share in this. But it is not a drive which – as many psychologists believe – makes us want to return to the warm seclusion of the womb which protected us before birth. That would not be trust, but anxiety. We would not go back into the original condition of our existence but forwards into the future of our destiny on earth.

A bird seeks no refuge. It builds its nest in trust towards future generations which will carry further what it passes on to them.

A bird's nest can hardly be compared with the dens and holes, the shelters and lairs of other animals, for a bird *builds* its nest: it forms it and equips it. It uses feathers and branches, grass and twigs, stones and clay, straw and mortar. There are birds which weave their nests and others which make them firm with saliva, which turns into a kind of mortar on drying. In trees and bushes, in clay and sand, in grass and on rocks, birds' nests are built. Birds first touch upon earth-existence in the nest. It is the support, the foundation, on which their existence rests. Without a nest they would have soared away into the widths of the air. But their house keeps them fixed to the earth; it is a placenta through which they remain living within the surroundings of the earth. But the ability to build a nest is at the same time the umbilical cord which binds them to it.

For other creatures the nest is a refuge and a shelter. With few exceptions (some fish or the beaver, for example) the nest is not built but searched for and found, or it can be dug and scooped out.

Sparrows give their name to by far the largest order of the bird world, consisting of over half of the world's bird population. They are the Passerines, with about 5,800 different species.

> The Passerines are spread over the whole earth and inhabit woods, heaths, moor, tundra, high mountains, steppes, deserts, reedy thickets and open, tree-covered country ... They are originally tree or shrub dwellers, as most of them still are today, yet there are some species, such as the larks and fallow-finches, which have adapted themselves completely to life on the ground. Some kinds ... are almost incapable of flight.[3]

And Brehm said of them:

> They are children of the land. As far as plant-growth

reaches, their living area extends. In woods they are more abundant than in areas without woods ... Many kinds live almost exclusively on the ground and most of them are not at all unaccustomed to it. Very few shun the vicinity of people; many invite themselves as guests to the master of the earth in so far as they trustfully visit his house and farm premises, his orchard and flower garden.[4]

Crows and jackdaws and magpies, finches and bullfinches, larks and tits, thrushes, wrens, nightingales, swallows and hedge-sparrows are members of this order. All are birds which are connected with human beings.

But among them – perhaps bound most deeply to the earth and to people – house-sparrows form a special breed, for no other bird comes so close to humans. One could almost say that they alienate themselves from nature – in the manner of domestic animals – in order to become part of the human kingdom. But sparrows do not forsake their own form, as do all the varied types of dog. Nor do they lose themselves in a special activity as do hens and cows in producing eggs and milk. Sparrows are not domestic creatures yet they live as close to humans as few other creatures do.

The architects' guild ought to carry the sparrow on their coats of arms. Streets and squares, courtyards and gardens have become a natural environment for sparrows. They build their nests where people live. Under the roof ridge, behind the gutter, in the shelter of the chimney, at the edge of dormer windows. But the nests are also set up in the branches of trees and bushes where these are close to people's houses.

When sparrows were introduced into America in the middle of the nineteenth century it was observed that they appeared in larger, then in smaller towns, after that in villages and hamlets, finally on single farms. But if they were brought to a farm near a township, in a district where there are not too many of them, they would leave the farm and withdraw to the township.[5]

Yes, they look for the proximity of many human dwellings. There, where dust and waste, rubbish and refuse accumulate, out of straw and paper, hay and wool, hair and threads, twigs and grass-stalks, the sparrow builds its nest. It is not a very skilfully made house; from the outside it often looks dishevelled and without plan. But inside it is smooth, well-lined and warm. A roof always arches over it so that it becomes a real hut. Only when tree hollows or breeding cages are at hand is the vaulted roof omitted.

The material for nest building is brought together from everywhere. Above all the inner bark of twigs and branches is torn off for this purpose, and sparrows have even been seen daring to approach doves in order to pluck out feathers for their nests. No wonder that one of their many names is 'thief' or 'robber'.

The nest, once built, is lived in and guarded at all seasons apart from the 'holiday' weeks. No other sparrow may approach it; females drive off females, and males the males.

Mating, egg laying and rearing take place in spring and summer. Around Easter the female lays an egg every morning for about a week. From two to eight eggs are laid, depending on the age of the parents and on whether the season is early or late. The average brooding time is twelve days, but longer and shorter periods have been observed. As with all Passerines, newborn sparrows are naked and helpless. They have to be fed and warmed by the parents. But in less than three weeks they have grown up into little sparrows; they leave the nest and begin to look after themselves. Their parents then bother no more about them, even refusing them access to the old nest, for by this time a new brood is on the way.

Each sparrow pair breeds at least twice a year; some manage three or four broods, so the rapid increase of sparrow population is not surprising. As long as the brood is still young the parents care for it with touching devotion. They ward off, to the point of self-sacrifice, any attack on the little ones. Many observations have established this.

161

But as soon as the young have learned to fly, it is the turn of the next generation to take up their whole attention. So between ten and twenty young ones are hatched out and brought up by one set of parents quite soon after their first mating. Only when the third or even fourth brood has grown up, after midsummer, do the parents go away into the countryside. They have accomplished their work and can now devote themselves to the joys of existence. Other parents appear and young sparrows begin to spread out among the older ones and to claim their own living-space. Soon they will set up nests for the winter and will find mates. In the following spring they will begin to breed.

How quickly all that goes! Nevertheless everything is ordered and ordained. It is said in the Gospel, 'Are not two sparrows sold for a penny? And not one of them will fall to the ground without your Father's will' (Matt.10:29).

Sparrows and human beings

During the last hundred years sparrows have started an astonishing conquest across the earth. In the middle of the nineteenth century they were unknown in North America. Towards the end of the 1860s they were brought to various East Coast cities, and in a few decades they spread across almost the whole continent. By the end of the century they had penetrated as far as San Francisco. In 1889 W.B. Barrow wrote: 'From this time (1875) to the present, the marvellous rapidity of the sparrow's multiplication, the surpassing swiftness of its extension, and the prodigious size of the area it has overspread are without parallel in the history of any bird.'[6] It was a triumphal procession of the sparrow from the east to the west of North America.

By the end of the nineteenth century South America – especially Argentina, Chile, Peru and Bolivia – was conquered in a similar way. In Australia and New Zealand the coastal districts were populated at the same time by all kinds of sparrows. Today one can truly say that (except for Antarctica) all continents and – with few exceptions – all countries and regions are densely colonised by house-sparrows and allied species.

Their adaptability is astonishing. They are not bound to definite climatic and environmental conditions, as are many other birds. They can maintain themselves in subarctic districts just as well as in tropical regions and deserts. They are as much at home in Kiruna in northern Sweden as in Cuiabá in central Brazil. They live and nest also in barren Aden and tropical Burma.

There is obviously an intimate connection between sparrows and human settlements. In America one could observe that they pushed forward with the construction crews along new railway lines that were being built, from one inhabited place to the next. They always favoured settlements which had streets and squares, not isolated houses.

When empty houses stand free for nest-building, they avoid this opportunity, for the house sparrow 'will normally only nest in unoccupied buildings, if they are close to inhabited ones.'[7]

During the Second World War sparrows moved with the British Eighth Army through the North African desert and settled in their camping places. In 1956 a whole colony of sparrows had settled deep under the earth in a coal-mine in Northumberland, England, and were being fed by the miners. But this intimate connection between people and sparrows is optional, since we find sparrows also living, for example, on the uninhabited islands around New Zealand and in the deserted plateaus of northern India.

In spite of these exceptions – or perhaps just because of them – the connection of these birds with the human race is so astonishing as to be of a quite special kind. What is it that links the sparrow to man? Is it human beings themselves, or something else connected with them?

Sparrows always remain shy and wary towards people. Very seldom do they become tame and trusting, and attempts to get them to breed in captivity have not worked well. They lay only a few eggs and go about the business of bringing up their young in a dispirited way. They remain always on their guard when someone approaches them.

They can indeed be cheeky and will intrude boldly into kitchens and living-rooms to fetch food that appeals to them. But they stay for seconds only and fly off as quickly as they came. They live near, but by no means *with* us.

It is rather as if they like to dwell in the shadow of our human activities. Where we have worked and built, where traffic rolls and machines operate – in railway halls and factory buildings, in streets and houses – there sparrows like to live. There, in the refuse of our activity, in dust and soot and smoke and sand, the sparrows find a world which suits them.

This cannot have been always so. For only in the last hundred years has this progressive industrialisation of human existence taken place. Where were the sparrows earlier? Their history and gradual development is unclear. Some believe that they have pushed forward gradually from Asia to Europe; others believe that their origin is to be traced in Africa, whence they advanced north along the Nile valley. The various kinds of sparrows (rock sparrows, Italian and Spanish sparrows) which intermingled in many parts of the earth, have concealed their previous history.

But one thing is clearly to be seen: the gradual spreading of sparrows occurred in earlier millennia along the paths of farming peoples. Wherever people settled down to till the land, the sparrows who had come with them settled too. They became the farmer's companion, without ever coming close to him.

Perhaps we may outline the following history of the sparrows on earth: first they went with the farmers and the bounty of the harvest. Then they accompanied people into larger settlements and towns; and finally – from the nineteenth century on – they have gone into the industrial world that people have built up.

The report from a Northumberland coal-mine indicates a further step. Accompanying man, the house-sparrows dip into the depths of earth-substance. Yet they remain an independent race that has not lost its merry artlessness, its liking for play and pleasure, its spontaneous chatter and its trust in the world. They carry their bird life into the depths of the earth and yet retain their own nature.

A Christmas story

Once the sparrows were like other wild birds. They built their nests in trees and bushes, lived in the open world of meadows and steppes, on mountains and in the plains where brooks and rivers flow. At that time they could still sing and – like larks and finches, tits and starlings – make their voices resound into the widths of space. In those long ago days they still went on great journeys, moving south in autumn and back to the north in spring.

Today young sparrows feed mainly on insects and only a little on grain. Older sparrows have given up meat and prefer grain. This indicates that a gradual transformation of feeding habits has taken place. In the days when sparrows were still a proud, free race, they ate what nature offered them: spiders and beetles, grasshoppers and caterpillars and many other little creatures.

But step by step, they have given up the freedom of the wild. Their singing changed into chirping, twittering and chattering. The great migrations were discontinued, for man had begun to settle on the earth. He broke up the ground with his plough and sowed the seeds of the grasses that were to be transformed into grain for bread. Then the various wild sparrows came, looked at this human activity and found it right and good.

They reported what they had seen to their guiding spirit, who carried it further into the heights, and the angels found out what was happening on earth. The news rose to yet higher spheres: 'Man is cultivating the earthly ground,' it sounded. 'He sows and harvests and prepares his bread.' It was a beautiful song that sounded through heaven.

Now to the guiding spirit of sparrows came back the call that their creatures should accompany man at his work. 'Put aside your colourful clothing; forego your song; go to live near the children of humankind. Let your food be the corn; your nests the roofs of human habitations.'

The sparrows on earth heard this call and their hearts began to beat faster. 'O what misery,' they twittered. 'O what joy,' chirped others. 'Yes, we will bear it and do it.'

Since then the sparrow's heart beats ten times more quickly than the human heart: 800 times a minute, 48,000 times an hour and over a million times a day. Their breathing also is rapid; 200 breaths a minute, when they hop or fly, or when it is very hot outside. And their body warmth is much higher than feverous temperature in humans. These little birds are unpretentious but warm, with their hearts beating so fast that we can no longer distinguish the single beats and normally cannot count them. For us they are like a light, anxious fluttering.

But for the sparrows it is joy for they are responding to their guiding spirits, who had heard the message of the heavenly choirs. But hidden in those words was a mysterious proclamation which neither sparrow nor sparrow-spirit could decipher. It sounded something like this: 'When people grind flour out of corn, when they bake bread out of flour and water, leaven and salt, and take it as nourishment, when the earth is transformed under the burden of the growing ears – then a Child shall be born who will carry the Light. To him should your bird creatures be of service.'

Although the divine message was not understood it was none the less heard, carried to earth and imprinted in the hearts, beaks and wings of the sparrow tribes. Now the sparrows go where human beings go – into the darkness of earth and the realms of decayed matter; into the dust and the shadows.

But with them goes a delicate ray of light, fine as a breath, which only children on Christmas Eve can sometimes see. For the sparrows have become poor; their garment is grey, like poverty, and brown like the homely bread of earth. But in this garment of contentment lives the cheerfulness of small joys and the blessing of trust.

'What harm can the darkness do us?' beats the sparrow heart. 'The message is fulfilled, the Child is born, and light breaks into the darkness of earth. Down to the dust and the refuse, down to the depths of the earth, it penetrates, and we,' so the sparrow tongues twitter, 'perceive its shining.'

Messengers of Christmas are the sparrows. They accompany human beings and yet remain strangers to them. But one day, when in human hearts also the light of Christmas begins to shine, they will become tame. Then they will sit on our shoulders and pick breadcrumbs from our hands. And some will hold these crumbs in their beaks and will place the crumbs of joy on the lips of human children. That will be a joyous message for many angels in heaven.

9. Dolphins – Children of the Sea

Contemporary interest

In recent years new interest for the strange life of the dolphin has awoken in many places. In newspapers and magazines one finds articles and an array of books depicts their nature and characteristics.

This curiosity has particularly been aroused by various American researchers, firstly at the Marineland Institute in Miami, for the past few years in St Thomas on the U.S. Virgin Islands, and in other places. There extensive studies have been made with a certain dolphin, *Tursiops truncatus,* the bottlenose dolphin. The researchers are of the opinion that these especially intelligent animals can be trained to develop a certain degree of understanding for language.

This research, especially that of John C. Lilly and his team received plenty of government funding because the hope was that these studies would be of use later as a basis for developing communication with beings from other planets that may be met during space exploration.

Therefore it is a strange mixture of utopian and scientific curiosity that is spurred on by the special charm that these animals have. Indeed, anyone who has seen and observed dolphins can hardly escape from their fascination. They convey a feeling of contentment and joy. Some years ago a dolphin appeared at Opononi, a small place on the North Island of New Zealand,

and made friends with the children and fishermen who lived there. Soon thousands of visitors came to watch the sociable animal at play:

> Some people got so excited when they saw Opo ... that
> they went into the water fully clothed, just to touch her
> ... In the evenings, when it was too chilly to be in the
> water any longer and the dolphin had gone off, everyone
> talked about her. In the tents ... the campers exchanged
> their scanty knowledge of the marvel, speaking in low
> voices while the children slept. They visited each other's
> tents, becoming friends with total strangers in an instant,
> all because of the dolphin ... There was such an overflow
> of these friendly feelings that it seemed the crowds were
> composed of people wanting to be forgiven.[1]

Similar stories are known from antiquity. Herodotus, Pliny, Phylarchos and many other Greek and Roman writers record remarkable meetings and experiences with dolphins, describing their friendship with children and young men, and their helpfulness and self-sacrifice in saving those who are drowning. Recently four Japanese fishermen, whose boat capsized about 50 km (30 miles) from the coast, were rescued by dolphins. Each carried two men on its back to the shore. This story was told by the fishermen themselves.[2]

In the legends of Greece and Rome there are many stories of dolphins; they appear on coins, cups, jars, grave-stones and mosaics. At no other time have they been so present to human consciousness.[3]

Aristotle wrote much about dolphins in his *History of the Animals* (Book 9, ch. 35):

> Among the sea-creatures, many stories are told about
> the dolphin, indicative of his gentle and kindly nature,
> and of manifestations of devoted affection for boys, in

and about Tarentum, Caria, and other places. The story goes that after a dolphin had been caught and wounded off the coast of Caria, a school of dolphins came into the harbour in response to his cries and stopped there until the fisherman let his captive go free; whereupon the school departed. A school of young dolphins is always, by way of protection, followed by a large one. On one occasion a school of dolphins, large and small, was seen, and two dolphins at a little distance appeared swimming in underneath a little dead dolphin when it was sinking, and supporting it on their backs.

From these and similar observations, it can be seen that these animals have not only a special relationship to human beings, but also live in close relationship and co-operation with one another.[4] It seems to be one of their characteristics that they show helpfulness to others; something rare among animals, which deserves close study. What are the dolphins, and what is their origin?

The whale family

Dolphins and porpoises all belong to a larger group of mammals which live in the seas and in some rivers. They are included in the order of whales *(Cetacea)*. The suborders are the toothed whales *(odontoceti),* and the baleen or whalebone whales *(mysticeti).*

The former have, as their name indicates, fully developed mandibles. The baleen whales, however, have no teeth; they have in their mouths great racks, in which the small creatures which they eat are caught.

Baleen whales are the largest animals now living, among them the blue whale and the bottlenose whale which are hunted in the oceans of the Arctic and Antarctic. They are mysterious creatures, whose behaviour and way of life are still obscure. Their migration, their appearance and disappearance are only partly known. For millennia they have been attacked and exploited by men; but their mystery has not yet been revealed. The story of Moby Dick tells of this battle for the unveiling of their well-guarded secret.

Very different are the *odontoceti*. Among them there are giants too, like the sperm whale and the white whale. Both of these are much exploited, for they are often not much smaller than the mighty baleen whales. But they all have teeth; the sperm whale, however, only on the lower jaw. There is also a strange group, the narwhals, which have generally only one gigantic tooth, in rare cases two, which projects from the upper jaw as a spiral horn two or three metres long. Otherwise they are toothless.

Several of the *odontaceti* are fierce predators which rend and kill any creature that they meet. There is the fearful killer whale, with a dorsal fin projecting upwards like a mighty dagger. A Danish biologist, Eschricht, in the nineteenth century found in one of these creatures which had been caught and killed, the still fresh bodies of thirteen porpoises and fourteen seals. The fifteenth seal had stuck in the throat of this monster and suffocated it.[5]

The harbour porpoise, *Phocaena*, lives particularly in the northern waters of the Atlantic and sometimes swims up the great rivers – the Rhine, the Elbe and the Thames. There seems to be no difficulty for the *odontoceti* to change from salt water to fresh. Even killers have been observed and caught in rivers.

Dolphins themselves are spread around the whole globe, but inhabit particularly the seas of the northern hemisphere. On the coasts of Australia and New Zealand the large bottle-nosed dolphins with which the present investigations in America are concerned are found.

This short and incomplete survey nevertheless gives a first picture of the whales. According to the present results of research it appears to be justified to regard the *mysticeti* as having their home more in the Arctic and Antarctic regions, and the *odontoceti* nearer to the equator. The river dolphins whose snouts are extended in a beak-like form are found particularly in equatorial countries, for example in the Ganges and the Brahmaputra, and related forms in the Amazon, Yangtse, Mekong, Rio de la Plata and the Orinoco. Only *odontoceti* live in the rivers, and these in the temperate and equatorial regions. The *mysticeti* inhabit mainly the seas round the Arctic and Antarctic circles. The *mysticeti* which have a longer history on earth live nearer to the polar zones, whereas the *odontaceti* appeared later and moved closer to the equator. The two orders overlap considerably in their geographical distribution; nevertheless this appears to be the general scheme.

The dolphin's life in water

One characteristic belongs to both groups: they are given up entirely to life in the water. Seals, *Pinnipedia*, still change from the sea to the land and back to the water again. These mate and breed on the land, and the calves have to learn to swim and hunt in water. But whales are entirely confined to the water; they do not return to land at any stage of their lives. Only if the sea throws them upon the coast, and they are unable to get back into the water because of injury or the retreating tide, are they connected indeed with the solid earth but this is then their grave. They are much more definitely creatures of the water than seals; in spite of their mammalian nature they have left dry land entirely. The seas and rivers have become their pastures.

Their bodies have become more closely fitted to life in the water than those of the seals. For example, they have no hindlegs. Only rudimentary pelvic bones have remained and prove that rear limbs once existed. The arms too have atrophied and changed into two fins, which are used for guiding their movements, not for propulsion. And most *odontoceti* have developed a large dorsal fin, which the seals lack. They have thus many characteristics in which they resemble fishes. But the tail has a fundamental difference. This is not vertical as in the real fishes, but horizontal with some detachment and freedom from the backbone. It impels the animal like a propeller with a quick and powerful rotating movement which its position makes possible. All forward movement comes about in this way. Dolphins easily achieve a speed of 55 km/h (35 mph).

In their movements these animals are great artists. They are not only fast swimmers, who can race with any large ocean steamer, but also tremendous jumpers and divers, often hopping over small sailing boats as if to tease their crews. They can leap for several metres out of the water and dive in again gracefully and easily. Their moving games are varied, and much more agile than those of the seals.

In their games they can balance balls on their snouts, throw motor tyres, and catch objects. In the Marineland Institute at Miami, Florida, dolphins can be observed playing with the feathers of water birds; they let these be carried away by the currents in the large tank, pursue them, and bring them back. They can carry human beings as riders on their backs through the water, showing both the speed and the steady balance of the movement. No fish could do this. All dolphins live in pods; they are very seldom to be found alone; even when they make friends with human beings, they are often accompanied by another dolphin. As has been mentioned, they help each other when one of them is wounded or in need.

Alpers tells how some years ago seven dolphins were stranded on a small island in the north of New Zealand. They were seven or eight feet long (2.3 m), and the holiday-makers (friends of Alpers, who told him what happened) made every effort to drag them back into the sea. This energetic rescue attempt was however a failure, because of the community feeling among the group; for directly one of the seven was in the water, he attempted with all his power to join his stranded companions. None of them was willing to leave the others in their need. After many hours of effort, two dolphins were rescued; the remainder died on the shore; their sense of community did not allow them to preserve their own lives when the others were losing theirs.

Behaviour of this kind is only found otherwise among birds and mammals when they have their young. Then the mothers sacrifice themselves to preserve the life of their offspring. For this to happen on behalf of every member of the pod, as with the dolphins, is unique.

Lilly reports the following occurrence:

> An animal that was being delivered to an oceanarium struck his head on the side of the pool as he was being let into it. He was knocked unconscious and dropped to the bottom. The other dolphins pushed him to the surface and held him there until he began to breathe again.[6]

There are many similar observations. A mother dolphin holds her offspring on the surface until it has drawn the first breath. For the birth takes place, as mating does, underwater.

We are thus dealing with mammals which are given up entirely to life in the water, but which always have to come to the surface to breathe. All whales, *odontoceti* as well as *mysticeti*, can only live in water if they can breathe air. This happens at very different intervals and under very different conditions. It is stated that dolphins come up at intervals of three to five minutes in order to breathe out used air and breathe in fresh air. They then breathe very quickly, exchanging each time five to ten litres (1–21/2 gal) of air.

The senses

From this description it is clear that the organ which enables air to be breathed in and out is, for dolphins and whales, the centre of their life. This is the blowhole at the top of the head. It consists of a small hole which can be opened and closed by a valve. The muscles of this complicated apparatus are arranged in such a way that it is opened actively, and closed passively. This hole is

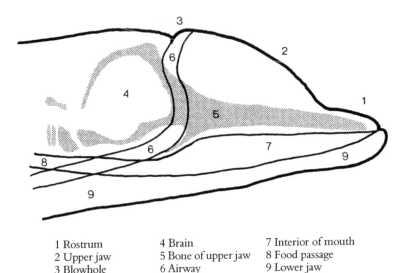

1 Rostrum	4 Brain	7 Interior of mouth
2 Upper jaw	5 Bone of upper jaw	8 Food passage
3 Blowhole	6 Airway	9 Lower jaw

Section through the head of a bottle-nosed dolphin (after Lilly)

like a pupil for the air, actively opened outside water and drawn together passively when the dolphin dives.

This central point at the top of the head should not be compared with the crown of the head in the human being. For only behind the blowhole does the bony cover of the skull begin, within which the brain is embedded. This place could only be compared in the human face with the region at the root of the nose. But the dolphin lacks the external nose entirely, and only remnants of the inner nasal passages are preserved in the neighbourhood of the blowhole. What appears like a nose is the upper jaw, cushioned with fat and oil. From the blowhole an airway leads down vertically through the upper jaw into the complicated larynx. From this the inhaled air proceeds through a short passage into the branching bronchial tubes of the lungs. The apparatus for closing the blowhole and the air passage leading to it is a system of valves and rings, which can be opened and closed independently at various places.

Immediately below the blowhole the air passage widens into

two larger and two smaller air reservoirs, arranged symmetrically. These air reservoirs, which surround the air passage and are enclosed in a thick net of muscles, serve for the production of sound, which can be achieved both under water and in the air. The larynx too not only regulates the stream of breath, but can probably produce sounds under water. It is now certain that dolphins can communicate through the most varied sounds, consisting of a deep creaking sound, and notes which extend over grunts and whistles into frequencies which are inaudible to us. These are emitted in the presence of danger and to call other dolphins to help. (The seven dolphins which were stranded, as was described above, communicated through these high frequencies.)

Further, dolphins emit another kind of supersonic wave: this works on the principle of echo location. Solid objects in the water – for example single fish, shoals, ships and rocks – echo or reflect these sound waves, which are then perceived as a sense impression by the dolphins.[7] These supersonic signals are sent intermittently and in varying rhythms:

> Apparently it [this sonar system] can be used to determine not only the distance and direction of an object but its form as well; by this technique they can find a fish that they want to eat and distinguish it from other objects.[8]

It is probable that the entire surface of the body is sensitive to these reflected sound waves. For the skin is specially fine and smooth, but nevertheless as if invulnerable. For when a wound is caused, from the lower fat layer an oil penetrates into it which stops the bleeding, closes and covers the opening.

No single hair is to be found on the skin of the dolphin, not even bristles or moustaches on the face. Everything is smooth and shining, perhaps in order to make the perception involved in echo location as clear as possible.

Through the special position of the blowhole and air passage the nose has atrophied. The entire loss of the organ of smell and the two olfactory nerves is connected with this. We have to suppose that no smell of any kind is observed by dolphins, when they draw breath on the surface of the water. It is still in doubt whether they have a sense of taste.[9]

But one sense organ is very highly developed; the ear. Although it is hardly visible externally, as its position is only indicated by a tiny opening behind the eye, and although the outer auditory canal is very long, narrow, and bent in the form of an S – the middle and inner ear are developed with great complexity. The bone which contains them is specially hard – the hardest of all bones found in animals. This is not, as with humans and most mammals, an organic unity with the skull, but only connected with it by ligaments and muscles. In this way it becomes possible for this ear bone (*petrosum*) to be moved in many directions and to adapt itself to a particular sound.

The real inner ear, the cochlea, is much larger in dolphins than in humans. It has two convolutions and is particularly developed in the lower part which is connected with the perception of high notes. The auditory nerves connected with this organ are also much more powerfully developed than in humans. If one considers that the capacity for perception for high and very high notes is very much greater in dolphins than in other mammals including people then the special development of the whole ear can be understood. Human beings can distinguish notes lying approximately between 20 hertz and 20 kilohertz. With whales and dolphins, but particularly with the latter, this sensitivity rises to 280 kilohertz.

Thus the dolphin, and probably most other whales, are a group of animals with a quite special capacity of hearing. They perceive their environment especially through sounds, tones and noises. Their realm is not resounding air but the reverberating water. They breathe air but they experience their watery environment as a sounding and rippling, as the resounding of the breakers, the steady murmur of the waves.

Is it astonishing that the Greeks associated the dolphin both with Dionysus blowing his flute and with Apollo, the master of music? That they let the dolphins play to the sound of the lyre?

The brain

Another special characteristic of all dolphins, porpoises, and other whales, is the exceptional development of their brain. No other mammal, not even the anthropoid apes, can be compared with them. In absolute and relative size of brain the *odontoceti* are the only mammals which are close to the human.

That such mighty creatures as the blue whale or the sperm whale should have enormous brains is not surprising. The brain of the sperm whale weighs about 8 kg (18 lb). But dolphins and porpoises that are not much larger than humans have about the same size brain, which is an extraordinary phenomenon.

In addition, the form, structure and convolutions of the dolphin brain are remarkably similar to that of the human brain. The former, indeed, appears as if compressed at the back; but it is not much behind the construction of our brain in the complexity of its development. The convolutions and folds are very numerous, and recent investigations have shown that the number of nerve cells is similar to that of the human brain substance. The same must be said of the cerebellum, in every respect. It is not only large, but complex and similar to the human cerebellum.

Among the nerves of the brain the olfactory nerve is missing, being only present in a rudimentary form in some baleen whales. The visual nerve too is relatively small. Only the auditory nerve is outstanding in size. It is much the most highly developed sense nerve, and in connection with it those parts of the cerebrum which are connected with hearing are particularly well developed. Here too is shown the nature of the dolphin which is so deeply concerned with the realm of sound.

But why is the brain so large, and so human in form? This has become a burning question to those who have occupied themselves with this matter. Lilly and other researchers in North America chose the bottle-nosed dolphin as the object of their experiments because they assumed that the size of brain should permit communication like that of speech. Lilly, for example, is convinced that the capacity of speech is bound to a specific size of brain and that the growing child can only learn to express itself through speech when the necessary size of brain has been reached around the second year.

Such considerations certainly deserve to be treated with a certain scepticism. But the question remains: why have dolphins and other *odontoceti* such a highly developed brain? Many researchers bring this into connection with their skilful capacity for movement; others believe that the particularly active metabolic system is responsible. But where there are such varied possibilities of explanation the real answer has not yet been found.

In this connection the astonishingly mature social behaviour of the dolphins is to be considered, which has been already described; the strong connection they have with one another and their amiability towards humans. It must be remembered that dolphins behave quite differently, for example, towards fish. It is known that fishermen in various regions are assisted by dolphins. They summon them with high-pitched whistles; and the dolphins drive shoals of fish towards the boats, as hunting dogs drive game towards hunters. Pliny describes this behaviour; so does Oppianus, a great admirer of the dolphins. In our times similar observations have been made.[10] Dolphins behave with particular enmity towards sharks. They attack them by driving their closed jaws into the body of the shark and tearing open the wound made in this way with their sharp teeth, of which they have eighty-eight. But a human being has almost never been chased or attacked.

Lilly's assertion that he has never seen a sleeping dolphin is important:

> Because they do not constantly have to resist gravity
> as we do, they do not need to sleep as we do. As we
> discovered, they cannot afford deep unconsciousness at
> all from any cause – anaesthesia, epileptic convulsions,
> or a blow on the head hard enough to produce
> unconsciousness will kill them.[11]

This can be understood if one remembers that all whales have to come to the surface to breathe at shorter or longer intervals. If this does not happen they are doomed. Is this perhaps the reason for the size of their brain – that it protects them from falling asleep, that different parts of the cerebral cortex can be active in turn, while parts that have been exhausted by consciousness can recover?[12]

How otherwise can this continuous consciousness be achieved? There are many riddles here waiting to be answered. But does not the ever-waking dolphin bring into the depths of the ocean an element which can be felt like an illumination of the water? Fishes have a dim, dream-like consciousness. Whales however carry their constant day-consciousness down into the depths of the ocean, bringing light into the darkness ruling there.

The essence of the dolphin

To bring all these phenomena into a single, inclusive and comprehensive picture, revealing the nature and development of the dolphins, is a difficult undertaking. Is it right yet to attempt any such image, tentatively or definitely?

We picture the oceans of the earth, through which there wander the mighty baleen whales; like memorials of a primeval age they plough the waters in the polar regions, grazing on the billions of small creatures which constitute their food. They are joined towards the tropics by the toothed whales; from the mighty sperm-whale to the fierce killers, and the schools of dolphins and porpoises. For all these, although they are mammals, life in the water has been prescribed by world destiny. They have continually to come up from the depths in order to breathe in and out, and this demands uninterrupted wakefulness; sleep and any other form of loss of consciousness bring death. Therefore their brain is specially large and has as complex folds as that of man. The nose is turned upwards, so that its opening, transformed into a blowhole, comes to lie at the top of the head.

In this way mouth and nose, which are otherwise among mammals closely connected, are anatomically separated. Between them an unoccupied space has remained which gives to all whales and particularly to dolphins their peculiar facial form. The widely spaced eyes survey this empty field of the upper jaw. High up breath is expelled and inhaled. The clear separation between nourishment and breathing is a particular characteristic of the whales. Through the blowhole which points upwards they have given to breathing a special place. For them it must be similar in character to the human capacity for forming mental pictures. For whales have no rhythmic breathing process, as have most of the other animals; their breathing is dependent upon consciousness, upon the varying circumstances of their lives, upon the struggle and the joyful game of existence.

Furthermore the sense of smell has been lost by them. Wherever the power of smell decreases, it is transformed into another capacity. Rudolf Steiner once described how the loss of that powerfully developed sense of smell which exists among the animals, leads to the development of human intellect.[13] Thus the human face is not extended forwards as in the face of many ruminants, beasts of prey, and apes, but directed steeply upwards and downwards, granting space to the brow and the chin.

But where did the transformed sense of smell go in the dolphin? Here it went through a metamorphosis into the tremendously developed hearing, which opens upon the infinite realm of sound waves in the air and the water. The dolphin is a creature listening attentively to the world, forming its picture of existence through these perceptions. Through the breath these tones, sounds and noises are transformed into conscious experiences, which then probably become memories.

The human being is perceived by the dolphin by the eye; this is evident from many of the descriptions given by Lilly and his fellow-workers. When the human individual with and through his eyes, gazes at the dolphin, it becomes tame and friendly. Does a memory then arise in it of old long-past times, when it was still itself on the way towards humanity? At the moment of such a meeting of eyes, do pictures of its evolution dawn upon its consciousness? So that it then takes children or youths upon its back and bears them through the waves, indicating that it wished to become what they now are, and gladly acknowledges their humanity? This belongs to the visual world of the dolphin.

With hearing, the dolphin lives in the realm of nature, not of the human. In this it hunts for fish, defeats that ancestral enemy, the shark, and darts through the water. But directly seamen and fishermen appear in their ships and boats, it becomes friendly, happy and tame. Then the eye enters their consciousness, with light and air as the dominant environment.

It is as if the dolphin lives in a world divided into two; in the heights of air and light and in the depths of sound and water. In one world, which air sustains, he encounters man. In the other sphere, in which he finds his nourishment, and his living space as an animal, he encounters other animals, which are enemies or companions.

Slijper relates that in the Marineland Institute dolphins were frightened by sounds between 300 and 400 kilocycles (about from low C to high A).[14]

Such observations show clearly that our hearing and that of the dolphin have an opposite background of feeling: what for us is music frightens them and drives them away; what for us is painful, like a high-pitched siren attracts them.

They are thus remote from us – and yet a part of us. But what was it that caused them to seek out, when they departed from evolution towards humanity, the ocean depths instead of the dry land?

A Greek legend tells of the younger Dionysus:

> ... how he was carried off, as he stood on the shore
> looking into the distance, by Etruscan pirates. They
> bound him to the mast; but the bonds 'fell from his
> hands and feet. He sat there smiling, with dark eyes'. A
> mighty vine grew up about the mast and sails, and the
> sweet fragrance of a noble wine filled the boat and made
> the crew and their captain drunk. Only the helmsman
> remained sober and recognised that he had a god on
> board. But Dionysus took the form of a lion threatening
> the seamen; in terror they leapt into the sea and were
> there transformed into dolphins swimming round
> the boat. The helmsman alone remained free of this
> metamorphosis. The god was revealed to him as son of
> Zeus and Semele.[15]

For the Greeks the origin of the dolphin is connected with the work of the younger Dionysus. He who cannot master the fragrance and power of wine – so they may have thought – becomes a dolphin. The one who remains conscious and upright, as the helmsman does, may remain in the realm of humanity. Rudolf Steiner once said about this Dionysus:

> For the macrocosmic counterpart of our present
> I-consciousness, with its intellectual civilisation, with all
> that derives from our reason, and from our ego generally,
> is in fact the second Dionysus, the son of Zeus and
> Semele ... [Pointing to the other legend, which tells of
> Dionysus' journey towards Asia] everywhere teaching
> men the arts of agriculture, the cultivation of the vine,
> and so on ... Every variety of intellectual civilisation stems
> from the journeys of the younger Dionysus.[16]

He who cannot make his step towards the consciousness of the 'I' is left behind and becomes a dolphin. This, although it is expressed in a mythological picture, is a key to the nature of the dolphin.

The human consciousness of the 'I', when it is achieved, can only develop because the brain becomes a mirror for thoughts and mental pictures. Rudolf Steiner indicates this when describing the creative powers which brought this about:

> When the ancient Greek was directing his feeling upon the microcosm, upon man, he called this element – coming from the earth and thus macrocosmic – this element which played a part in the constructing of the brain, the *Dionysian principle*; so that it is Dionysus who works in us to make our bodily organism into a mirror of our spiritual life.[17]

Here the historic sacrifice of the dolphin can be clearly recognised. It throws itself down into the sea from the ship of human evolution which is led by Dionysus.. It leaves the vehicle of the developing intellect in order to take with it those forces of the depths which would otherwise prevent the body from becoming the mirror of human thought. The true children of Dionysus remain above, in the light; the powers of the depths, which the god only summons at definite times of year (at the festivals of Dionysus) remain with the whales, which once liberated humanity from them.

When the sound of the flute rang out, when Marsyas arose against Apollo and the forces which opposed Dionysus were let loose, then those powers worked, which were indicated in Greece with the word *delphos,* the 'womb'. These are the same powers which the Greek knew to be guarded by the power of Apollo at his central sanctuary, at Delphi, which bears the same name.

Mythology

The foundation of this sanctuary is described by two legends, among many others; both are concerned with the dolphin. One ascribes the origin of the oracle to Eucadio, the son of Apollo. He and his companion, the nymph Lycia, were shipwrecked. In their need a dolphin came and took both upon his back, carrying them to the foot of Parnassus. There Eucadio dedicated to his divine father the sanctuary of Delphi.

But it is recorded of Apollo himself that he once took the form of a mighty dolphin which lay down upon the deck of a Cretan boat which was sailing to Greece. By his mighty presence he compelled the boat to take its course to Crissa, the harbour of Delphi. Here Apollo sprang ashore like 'a star at midday. Many shining sparks flew from him, and a glory of light reached the heavens.' But Apollo appeared in the form of a youth to the terrified and wondering Cretans and led them to the sanctuary and consecrated them as the first priests of Delphi.

In both these myths, a dolphin led to the foundation of the oracle at Delphi. It became a special sanctuary because powers of the depths – incorporated in the Pythia, working through the priests – were there at work, though bound and mastered. Were these the same forces which were once conquered by Dionysus? Or were they still mightier forces of darkness?

An ancient legend speaks of two different dragons dwelling at the foot of Mount Parnassus: one a male named Typhon, the other female, named Delphyne. She was regarded as the greatest enemy of Apollo. In order to overcome this dragon, Apollo had to transform himself into it. This was the power that ruled over forces of nature which preside over the blind processes of self-reproduction and birth, in their continual repetitions. (For *delphos* is the womb.) From this 'dolphin' form Apollo rises free, like a star, becoming its master. Thus through the Pythoness the otherwise unbridled reproductive powers of the feminine nature, overcome by the sun-god, were able to speak. Thus the image of

Apollo lives on in the consciousness of the Greeks and of later peoples as slayer of the dragon. Rudolf Steiner described this:

> And the Greeks imagined Apollo as shooting his arrows at the dragon, as it rose from the cleft in the form of turbulent vapours. Here, in the Greek Apollo, we see an earthly reflection of St George, shooting his arrows at the dragon. And when Apollo had overcome the dragon, the Python, a temple was built, and instead of the dragon we see how the vapours entered the soul of the Pythia, and how the Greeks imagined that Apollo lived in these swirling dragon-vapours and prophesied to them through the oracle, through the lips of the Pythia.[18]

Apollo had to transform himself into 'Delphyne'. In this way he won the power to conquer the Python. In its vapours his being could then work. Thus Apollo bore, among many other names that of the dolphin: Apollo Delphynios.

In connection with the form of the sun-god we meet the imagination of the dolphin in its macrocosmic being. Here are these primeval evolutionary powers which are mastered by Apollo. The dolphins which accompany the younger Dionysus represent the same forces in a human, microcosmic form. In the wide spaces of creation the light-god Apollo conquers the dark forces of the depths. Within the human soul Dionysus masters the dolphin and thus makes possible the awakening of intellectual consciousness.

The two paths of initiation which existed among the Greeks are indicated in this way. One led the pupil out into the wide spaces of nature; the other into the depths of his own being. But at Delphi both paths were united. In spring Apollo came from the north and dwelt for nine months at the sanctuary. In winter Dionysus replaced him. Both were guardians of the Delphic, sub-earthly powers which are at work in the turbulent being and becoming of Nature.

Here we find ourselves at the centre of the mysterious destiny of our dolphin, which we set out to seek. The mighty baleen whales appear as the result of the Apollonian, the more southern toothed whales as the result of the Dionysian, activities in the cosmos and in man.

Now we can ask about the external evolution of both families. Can anything yet be known about this? Only in recent years have decisive discoveries been made in this direction. Through serological research it has been proved that the whales are very closely related to the artiodactyls (pigs, camels, ruminants). Both great groups can be traced back hypothetically to a common ancestor in the Eocene period, that is in early Atlantis. A few specimens of this family, which must have had many branches and forms, have been found during the last hundred years or so. They are included in the animal system under the name *Archaeocetes*.

They all have a dragon-like body. The limbs are atrophied, particularly the hind limbs. The bodies are of tremendous size. The nose still lies at the front and is closely connected with the mouth, as in terrestrial mammals. From this original form the ungulates as well as the whales are derived.

Should we not assume that this dragon-like mammal lived in marshes? Why otherwise would its limbs have atrophied? It waded and crawled; it could not walk. With the gradual hardening of the earth two distinct groups formed. One mounted the dry land and developed into the ungulates; the other remained in the water and developed into the whales. The whales have never been terrestrial mammals. They have passed through a metamorphosis from the dragon-like archaeocetes descending into river and seas. This came about in the course of the Atlantean age. This dragon transformation is quite a different process from that which led to the extinction of the great saurians at the end of the Lemurian period. For here a dragon – in which the lower nature of existence and of man is expressed – is metamorphosed into higher forms. The powers of light and of the sun, Apollo's deeds and sufferings, conquer him. This process of evolution appears to have reached its end towards the conclusion of the Atlantean period.

Into this period falls an event, which is described in anthroposophy as the third pre-earthly deed of Christ. At this time the Christ-being, working from beyond the earth, brought about a harmonising of the three powers of the human soul; thinking, feeling and willing are brought into equilibrium. Rudolf Steiner says:

> [The Christ being was] able to drive out from thinking, feeling and willing the element which would have raged within them as a dragon and brought them into chaos.
>
> A reminiscence of this is preserved in all the pictures of St George vanquishing the Dragon which are found in the records of human culture.

The Greeks beheld their god Apollo in the same picture.

There is a wonderful consequence of the harmony brought about in this way. '... a weak echo of it could be heard in the musical art cultivated by the Greeks under the protection of Apollo.'[19]

Here we meet again the picture found in many Greek myths: Arion, the singer and harpist, who returns to his home Lesbos, the place of Apollo's birth, carried by dolphins. Humankind receives the art of music through the sacrifice once brought by the dolphins. They took the powers of the dragon with them into the depths of the water; thereby something else was rescued.

The human path led upward; humankind was accompanied by the race of ungulates. Beneath their feet the marshy earth that they were leaving behind grew harder. Feet became hooves. The herds could appear on the firm ground of the steppes and meadows that came into being. They grew teats and became givers of milk. Horns and antlers sprang from their heads; new, otherwise unknown symbols of their destiny. They raised these structures, like archetypes of those musical instruments which give men the art of strings, into the atmosphere. The cythera and the lyre appeared in their perfect beauty upon their brows.

What the whales keep as a mighty brain mass within their skulls is shown here outwardly in the form of horn and antler. In the human being both form and mass are transformed into the power of thought. Substance and shape are spiritualised.

Thankfulness that this could happen should fill our hearts with endless humility, when we look at the whale and the dolphin, at the ox and the sheep and the antelope. For this reason the ox could be one of the first to greet the Child lying in the manger. The ox could be present as representative of all the ungulates and the whales of the earth, at the place of holiest poverty.

10. Cats and Dogs – Human Companions

The nature of cats and dogs

Since the most distant times the dog has accompanied the human being. The two are so closely bound up with one another that it is fair to ask if it was ever otherwise – if there was some earlier time when man and dog lived separately. This is difficult to decide, for the paleontological prehistory of the dog still poses great riddles. It presents us with so many layers and different forms that a clear-cut solution seems hardly possible.

People still search for the one and only ancestral form out of which the almost countless varieties of modern and earlier canine types might have proceeded by way of breeding and selection. However, all the representations which have come down to us from former times – on Egyptian tablets, Babylonian seals and Greek vases – show the different kinds of dogs already familiar to us, as well as some forgotten ones: large and small, sharp and short muzzled, single-coloured, multi-coloured, spotted, long-haired and short-haired, with large ears and with small ears. Thus if we look back over three or more thousands of years, we already seem to recognise Pomeranian and setter, sheepdog and greyhound, pug and spaniel.

Moreover, paleontology has uncovered an equal variety of skeletons, which allow us to postulate the existence of most species of dogs even in the earliest Stone Age. In this area, the

researches of Theophil Studer are particularly significant.[1] He attempted a first ordering of the material available at the end of the nineteenth century; and though his findings are in part questioned today, his fundamental conclusions are indicative. He distinguished two basic forms: a northern, palearctic dog and a southern type extending down to India. Both accompany man and never occur as wild forms. It is generally believed today that continual crossing of dogs both with wolves and with jackals has given rise to the panoply of forms found in the great dog-families all over the earth. Indeed, it is characteristic of dogs and wolves that they have an amazingly mutable body, able to adapt to new conditions in the shortest time. For example, young wolves which have been domesticated from birth have shorter skulls and snouts than their wild brothers. Moreover, the astonishing variability among the dogs themselves – from the tiny Pekinese to the giant wolf-hound – reveals the plasticity of their form. Certainly no other animal can compare with the dog in this respect.

The dog is called 'the oldest domestic animal' because in early history it was already connected with its master. In an old Persian law-book it is said that the whole world exists only by virtue of the dog's intelligence.[2] Its power reaches from heaven down to hell, for below, in the realm of darkness, it guards the entrance to the underworld in the form of Cerberus. The Indians, for their part, worship Yama, God of the dead:

> For them, he is the first man who found, in death, the way to the bright heights for many born after him and there he reigns as the gatherer of humankind. His dogs keep watch at the entrance to heaven.[3]

Thus the dog stands at the threshold of death which the human being must cross. This helps us to understand why we find on many English and French gravestones a deceased knight portrayed with the soles of his feet resting on the body of a dog.

However, is it really correct to call the dog merely a domestic animal, and thus put him alongside cow and sheep, chicken, pig and horse? Does it not occupy a quite different position from theirs? For all other domestic animals we build stalls, maintain meadow and pasture land, and keep them near to us. The dog – and the cat – share house and habitation with us, even if sometimes we give them a little house or a basket of their own. These two – dog and cat – are much closer to us than cow and sheep ever were. The position of the horse is a special one; it will be discussed in a separate chapter.

We see the dog and cat accompanying people from very early times, in such an intimate relation that no other animal can be compared with them in this respect. There are animals near to humans who are unable to live without them. Many birds, such as the ravens and crows, doves and swallows, owls, swans and storks, and most of all the sparrows, seek out the vicinity of humankind. Bears, too, as well as elephants, hyenas and buffalo are very close to us. But they all have retained their own living space; they live *next to* man, not *with* him. The rule of nature and its seasons is still deeply bound up with their own being; animal and environment fit one another like key and lock.

With domestic animals it is different. Through domestication they have been divested of their natural connections. They need stalls and stables, enclosures and meadows, because they are no longer able to provide shelter and food for themselves. Struggle, lust for life, and wild roving have been transformed into service, docility and security. Now milk flows at almost all times; eggs are laid and wool given. Once near to us, they have come to serve people.

Dogs and cats, however, are not servants; hence they cannot be ranked among the domestic animals – though the literal meaning of the word suits them alone, for it is only they who sit by our fireplaces and stoves; they use our rooms and have their place at our side.*

Was there a time, long ago, when they had to be broken in and tamed? Or is this just a notion arising out of Darwinism? If falcons and wild animals are tamed today, this is not comparable to the process which led to the domestication of various animals. The young of tamed animals will never be tame solely because their parents have learnt submission. They will remain wild animals. What once must have taken place – assuming that the accepted view of domestication can stand – was so profound a change in the animal's essential nature that it passed on acquired manners and habits to its descendants. This thought alone is hard enough to understand or demonstrate experimentally.

Would it not be much more natural to think of a time when animals and humans were so closely akin that they lived with one another? According to this view *all* animals were once near to humankind; and only gradually did the wild forms arise, breaking away from the human world and assimilating themselves to natural surroundings foreign to man. In other words, wild animals were not gradually brought into a domesticated state; they *became* wild. Thus we can imagine a twofold path: one leading away from people towards wildness, and another, much shorter path, whereby the cattle and swine, sheep and goats become servers and providers.

In the light of this new picture we can see the evolution of dog and cat at our side in a much more immediate way. The two remain faithful to the eternal archetypes of their group: they have always lived at our side. It was only *from* them that the lion and tiger, puma, panther and leopard developed later – as well as the wolves, foxes, jackals and dingoes. Thus we must break through to a new image: the human being in the centre, accompanied by dog and cat. The dog trots on his left side, the cat on his right. Indeed, we may not assume that they have always looked the same as they do now. They were originally much closer to their

* It should not be objected here that the parrot, magpie, canary and budgerigar also live with us. Indeed they do, but only as captives; the must be kept tied or caged. Dogs and cats, on the other hand, seldom try to run wild.

archetypes; they were more changeable in form, not so special-ised. Nevertheless, the dogs and cats of today still stand closer to their primeval brothers and sisters than do the recent wild forms. The latter arose out of the 'generalised' dog and 'unspecialised' cat who accompany humankind out of prehistoric times on to the stage of history.

It is now time to characterise dog and cat as two prime repre-sentatives of the carnivores.

Characteristic traits of cats and dogs

It is not entirely simple to grasp the essential differences between the dog-like and the cat-like carnivores. And yet the differences are so obvious that there will seldom be real doubt as to which group a creature belongs. Where do these differences lie? They are found not only in the anatomy, but in the entire behaviour, temperament and manner of reaction in the two groups.

In the build of the body there are no fundamental differences. Both groups represent typical carnivores: swift, mobile, with strong limbs whose structure – even in the very jaws – points to the pursuit and attack of prey. The legs carry the light and often stretched-out body at a good speed, while the body offers no hindrance to run-ning, walking or climbing through its weight or unwieldiness.

Here, however, we meet the first differences between the two groups. The dog-like creatures have longer limbs and their body rises higher from the ground than that of the cats. Their legs are not only longer, but also thinner and more disposed for running. Not many dogs can climb trees, but to most cats no tree or shrub presents an obstacle. Jaguar and leopard leap on their prey from out of the branches, and carry their booty up to the crown of the tree where they can keep it safe.

Moreover, dogs do not have the kind of paws with which the cat tribe is endowed. A dog will hardly ever strike with his forefeet, nor will a wolf, a jackal or a fox. Cats, in contrast – and the lion and tiger too – strike out with their paws; and often the first attack is executed with a slap of the forepaw. Only then do the jaws set to work. Dogs are attackers only with jaws and teeth. They chase their game till it is tired, then set to and maltreat it with their jaws. Cats, however, creep up and pounce on it, striking it senseless with their paw. There are, of course, the cheetahs, who can hunt after a dog-like fashion. Hilzheimer says of them that:

> They unite the canine type with the feline form in a unique way ... They are suited for swift running, and thus have become quite long-legged and somewhat dog-like creatures.[4]

Two additional features are connected with this behavioural difference. A cat's claws are moveable; in running they are retracted, in striking and lashing they are extended to become a weapon. A dog's claws are blunt and immobile; but its mouth, in recompense, is much longer and contains more teeth. The felines generally have 30 teeth; dogs have 42. The two groups have the following dentition:

1	3	1	3	3	1	3	1		2	4	1	3	3	1	4	2
1	2	1	3	3	1	2	1		3	4	1	3	3	1	4	3

cats   dogs

Thus the dog's mouth is drawn out and accordingly holds more teeth; but the front legs lack the power to strike and tear. Among the cats this is reversed: the muzzle is much shorter, while the front limbs move more freely and are more independent of the pull of the earth's gravity. Thus, what in the dog is confined to the mouth is spread over the whole body in the cat.

A dog attacks with his mouth and teeth alone; partly for this reason his mouth has come to be a sort of ventilation organ as well. When a dog is overheated from exertion, he pants, expelling hot air and drawing in cool air with the breath.

Besides this, the nose and sense of smell are particularly well developed in the dog. He 'recognises' and 'remembers' in the realm of smell. This is his field of perception and orientation *par excellence*. It appears that the great attachment which the dog forms to an individual human being comes about through the sense of smell. Hilzheimer writes that after destruction of the olfactory nerves or loss of the sense of smell, this attachment is no longer evident.[5]

Felines are in general far more independent of human beings than are the dogs. They live in our vicinity but remain much more within their own world, even if they share our houses. These creatures are endowed, to a certain measure, with a self-contained and self-sufficient being. Wherever they may be, they carry their own rhythm and their own existence with them. Heedless of their surroundings, they lick and clean themselves; they withdraw into themselves if they so please, and seldom feel disturbed. Dogs, on the other hand, are part of what is around them. They live from the look and gesture of their owners; the world, and especially human beings, are their masters.

With this character-sketch we have brought out something of the basic type of each family. For the dog, the focal point of existence is its head. Muzzle, teeth and nose are exceptionally alert and sensitive. The ears prick up at once, the mouth opens slightly and the eyes are directed on the object which makes the smell or sound. With a jerk the head lifts up and draws the body after it.

Cats display a much more extended sensitivity. Their coat is uncommonly sensitive, and the whiskers around mouth and nose, forelegs and ear give them an all-round capacity for refined sensation. No doubt the most gentle breath of air, unnoticeable to other animals, will register on it. This is why a cat so dislikes getting wet, for it then loses its sensitivity, and this it cannot bear. Hence the majority of the cat family are extraordinarily shy of water. Dogs are quite different in this respect: they mostly like the water and are natural swimmers.

In this way another well-defined difference appears. In dogs, sensation is concentrated in the head; in the cats it is spread over the body. In cats we find the most delicate sense of touch, in dogs the highly differentiated sense of smell. Cats live in the interplay of sun and shadow through the air; dogs, more in the cool and watery elements. Is that why all dogs – as long as they are well – have a cool, damp nose?

The general impression of the feline form tends towards the round, the spherical; the canine towards the stretched and pointed. How angular and hard their gait is when compared with the rounded, fluid, soft and sometimes stealthy motion of cats! The dog storms ahead, and swiftly turning, he charges back, only to rush forward again. The spring, walk, and run of the cat are measured and self-contained. Each movement is directed and distinctly harmonious. The gait of the great cats is purposeful and coordinated.

If we wish to reduce this difference to a formula, we might say that the movement of the cats is controlled from the domain of the heart, while the dogs are at the mercy of their senses, so that their activity is constantly subject to changing impressions and sensations. Both groups are fundamentally orientated around the rhythmic system. The dog's rhythm is that of breathing – giving out and taking in, ceaselessly in communication with the world. Its head is the watchman over this respiration, guiding it, calling it, and also hindering it. The legs are long because the breath-stream flows into them and stretches them out. In running and chasing, the breathing determines the movement activity.

In cats, it is the heart which sets the rhythm and simultaneously strikes the fundamental note of all other movement melodies. It is the heart that cries after blood, but even in greed cats show moderation. The great cats will kill their prey, suck and eat, but will leave the greater part for others to consume. Then hyenas and vultures, jackals and wild dogs take over and feed on the remainder.

Sun-warmed air, wafted through a dry atmosphere, throws its illumination into deep forest; thorny bushes, short grass, sand, noontime repose or evening quiet: this is the world of the great cats. Their coats are yellow and brown, with stripes and dots, and peculiar spots that appear as though painted by the interplay of light and dark.

In the pale grey of evening and night, in fog and rising moisture, in caves and borrows, in the depths of the woods, where environment calls for constant wakefulness – here is the wild member of the dog family. Foxes and wolves seldom appear until twilight. They roam around, filled with fear and unrest; the gather – particularly the wolves, seldom the fox – in packs, and run howling through the darkness. Their coat of fur is long but of a single colour – pale and dull. They follow their sense of smell, and live in the realm of the slightly moist aromas given off by other creatures. Their species are not so numerous as those of the cats, but – because they are closer to the watery element – they have retained an extraordinary plasticity and mutability of form within the framework of their archetype.

Manes and coats

If the images we have outlined are at all true to the archetypes of dog and cat, they should cast light on the various species found in these two groups of animals. The living types should find their places within the limits of variation of these chief properties, for differences within an order or sub-order are only a stronger manifestation of one or other characteristic feature. All cats have the characteristic features of their family, but the single breeds develop one or other feature – or a related set of them – in a slightly one-sided way. This gives the form its individual stamp.

Characteristic of the lion, for example, is the marked difference between the sexes. Most males of the African, Near-Eastern and Indian lions bear an imposing mane. It consists of a great crown of hair surrounding the powerful neck, framing the face and often continuing down the back and belly. Guggisberg writes:

None of the other great cats shows such a well-defined sex differentiation as the lion. Any child can tell the maned lion from the lioness at first glance; with the tiger, leopard and jaguar, it is harder to distinguish the gender by casual observation.[6]

The coats of the other great cats are imprinted with unique patterns which the lion lacks. His coat is uniformly sand-coloured, without stripes or spots. Can we see these opposite features as a kind of polarity? What is the significance of the peculiar patterns on the coats of all the great cats, and on the little wildcats and house-cats too? Or of the tiger's dark stripes, the rings and spots of the jaguar, the light and dark patches covering the whole body of the leopard?

The small cats all display designs characteristic of their breed. Spots, stripes, rings, dots alternate with one another according to colour and pattern. The snow-leopard, ocelot, wildcat and many other forms bear such decorations.

What is the significance of this? In Novalis' *Disciples at Saïs is* we read:

Manifold are the ways men go. Whoever follows them
will see wondrous patterns arising; figures which seem to
belong to that great script we see encoded everywhere, on
wings, eggshells, in cloud, in the snow.

Is it the same 'code' that declares itself, so plainly written, on
the coats of the cat tribe?

Portmann draws a very apt comparison when he indicates that
morphological and biochemical investigation of such pigmentation will not lead far by themselves for the patterns and designs
are like an unknown script:

> Now the question arises as to the meaning, the
> interpretation of the word: and this must be solved with
> totally new means of investigation. The meaning of
> signs is independent of the materials with which they
> are written. Thus, for the question of the meaning of
> the pattern on an animal's head, the composition of the
> markings is of little importance, but the character of its
> configuration plays a great role. To answer this question,
> we must follow quite different avenues of study from
> those of physiological or genetic research.[7]

These avenues must be explored step by step, to gain an
understanding of the rune-formations of skin, spine, fur, wing
and feathers of living creatures. It is the 'great script' of Novalis
that must be deciphered.

Could we not begin this quest with the cat family, and take
the polarity 'mane versus coat-pattern' as our starting-point? The
lion's mane, usually darker than the coat, not only frames his face
but sets off the head morphologically from the rest of the body.
This results in a figure of which Eugen Kolisko once said, with
a wry smile, 'the lion doesn't fulfil behind what he promises in
front.' This captures the impression perfectly.

Without the mane, head and body flow together harmoniously; and what is otherwise held back in the formation of the mane now comes out in the patterning and ornamentation of the coat. Here we encounter a rule that in essence was indicated by Goethe. He drew attention to 'that idea of an economical give-and-take' which reigns in nature. 'Here we are at once faced with the law that no part can be added to without something being taken away from another part in return.'[8] The process which leads to an excessive mane in the male lion unfolds as colour and form in the other members of the cat family.

Is it the transition from head to body that is revealed by the coat markings? Can we comprehend the brown polygon patches in this way that cover the body of a giraffe from its over-long neck right down its long legs? And does the zebra have those dark stripes because it has such a strong neck? Can this lead us on the right track to grasping the inherent laws of formation?

Observing the great cats it becomes apparent that markings on a tiger seems to correspond to the sequence of the ribs. The black stripes stretch like bands around the breast and stomach, continuing similarly along the back legs and tail. Servals and ocelots (dwarf leopards) show dark patches and rosettes clustered in a striped pattern following the spine then along the legs and tail. The clouded leopard, like the giraffe, is evenly covered by dark patches. In this way we can distinguish three kinds of markings:

1) *Rings* encircling body and limbs. Examples: tiger, wildcat, striped hyena, zebra, okapi, bongo, Tasmanian pouched wolf, and so on.

2) Longitudinally arranged *spots, speckles and rosettes* running parallel to the spine and the long tubular bones of the limbs. Examples: serval, ocelot, Spanish lynx, genet.

3) A network of *polygonal and round patches* evenly covering the body: cheetah, snow leopard, giraffe, and so on. (Leopard and jaguar fall between groups 2 and 3.)

By arranging things in this way, perhaps through this threefold system we arrive at a preliminary indication of the meaning of this runic script. At least we shall not be far wrong if we imagine the power of the rhythmic flow of blood and breath to be the artist who designs these patterns. Breath, flowing in and out, draws the dark and lighter rings around the body and limbs of tiger and zebra. Blood, flowing to and from the heart, uniting body and head, paints the spots and speckles lengthwise over the body. And the network of patches on jaguar and giraffe is an externalised picture of the capillary blood circulation on the surface of the body.

Thus it could be the rhythmic organisation of blood circulation and respiration, so exceptionally well developed among the carnivores, that manifests itself outwardly here in design and pattern. It could be the task of a future animal psychologist to look into such questions; for the behaviour of tiger, leopard and jaguar also differs in accordance with these runes of their being.

Why, then, does the lion have a mane, thus remaining one-coloured; and why is this mane the male characteristic, while the lioness is both maneless and unornamented? Among all the great cats, the lion alone has remained one-coloured; and does it not share this feature – allowing for a few exceptions – with the dog family? Almost all the wild members of this family are single-coloured, and their hair is as a rule longer and thicker than that of the cats, approaching a mane formation.

Does this imply that the lion has a certain similarity to the canine family? But surely there cannot be a more typical cat than the lion itself! Or does it achieve its position as 'king' of the beasts precisely because, of all the cats, it alone manifests masculinity?

It begins to appear that we are observing formative tendencies which refer back to Goethe's 'economical law of nature.' When stripes, rings, and other patterns appear on the body-covering, mane and long fur are absent. On the other hand we meet uniform colouring where coat and mane grow full.

Thus it seems that all colours, spots and markings – along with short fur – represent feminine characteristics among the mammals. Longer hair growth, mane, and dull colour without patterns or markings are, by contrast, more masculine features. The mane, the lighter ring of fur around the wolf's neck, set off the head from the rest of the body. This too is a feature of masculinisation, while the greater unity of head and body, the merging together of these two parts, is more feminine.

So, as a preliminary sketch we may say: the cat-like animals are in many of their bodily characteristics of a more feminine nature. The dog family, by contrast, shows strong masculine traits. Only the lion steps out of this ordering to become a masculine member of the felines. Through his mane, he gives us a key to the morphology of the carnivores.

Masculine dogs and feminine cats

With these thoughts, we have come to another important feature of cats and dogs: colouration is simple and sparing in the dogs, striking and vivid in the cats – except the lions. Now we must attempt to connect the characteristics mentioned in the second section with this further quality.

The rounder, more self-contained form of the cats – both great and small – is linked with a significant aggressive power. Tiger, jaguar and panther are tamed only with the greatest difficulty; the small cats too are wild, aggressive, and eager for prey. Lions, however – especially males – are lethargic, slow, and often seem to have an air of apathetic superiority.

The dog family, also, is not given to immediate aggression. Foxes, jackals and wolves are not attackers. They take what prey is due to them and no more. The descriptions by the Crislers of their experiences while living with young Alaskan wolves are filled with a touching understanding for these animals.[9] When a wolf has learnt to trust you, it will show an astonishing degree of charm, sympathy, and attachment. Their wildness can suddenly break out again but it is not so general or consistent as with the cats.

Thus we can point to a parallel between the patterning of the coat and the behavioural character of particular animals. The wild, aggressive cats are characterised by a colourful, often vividly patterned coat. The dogs, who are much more peaceable by nature, show a simpler and more monotonous body covering. The only strongly spotted wild dogs are highly aggressive predators: the hyenas. Hence we can take it that among the carnivores, a higher degree of inner excitability is manifested in colouring and marking of the coat.

In the case of certain birds, Suchantke has pointed to similar phenomena.[10] Here as well there are clear connections between marking, colour, ostentation and the emotional disposition of a particular species. The simpler a bird's plumage, the more pronounced is its capacity for caring for the young.

However, when we turn to the formal principle in the morphology of dogs and cats, it is not so easy to grasp the connection between bodily build and character. The roundness and self-contained form of the cat's body would speak for greater gentleness; the angular forms of the wolves and foxes, the larger snout, more numerous teeth and quicker mobility would point to ferocity and greed for prey. Is this not a contradiction? Or do form and structure have converse significance from that of colour and marking? This is a problem that is also reflected in our speech habits. We often speak of the dog as 'he' and of the cat as 'she.' Here the wisdom of language expresses something that we perceive half-consciously. We cannot but experience the cat as feminine and the dog as masculine. Thus we can speak of the round, feminine forms of the cats and the more masculine, sharp forms of dogs. However, this flatly contradicts the manner of their behaviour.

This contradiction remains unsolved until we consider what Rudolf Steiner said in relation to the human sexes. In a lecture on 'Man and Woman in the Light of Anthroposophy,' he pointed to the contradictory ways in which the sexes were described around the beginning of the twentieth century. Some found the woman's character to be basically choleric; another gave humility as her dominant quality. Yet:

> Another researcher ... concludes that woman's essential nature could be described best with the word *devotion*; another prefers *the desire for domination*; another *conservatism*; and yet another finds that woman is the genuine revolutionary element in the world.

The contradictions found here are not unlike those we met in our examination of dog and cat. Rudolf Steiner takes up these contradictions and shows that they are quite legitimate; we must simply learn to see through them. First of all he speaks of the fourfold human being, who besides the physical, holds an etheric or 'formative-forces' body within himself, and, in addition, the soul body and the I. Then he says:

To begin with, only the physical and the etheric body concern us here, and this is where the solution to the riddle of the relation of the sexes lies hidden ... The being of man is a peculiar organism, for the etheric body is only partially an image of the physical body. In respect to sexuality, things are different. In men the etheric body is female; in women the etheric body is male. Strange as it may seem at first, deeper examination must lead to this highly significant fact: concealed within each human being lies something of the opposite sex.[11]

Suddenly, now, light is thrown on part of the mystery surrounding the dog and cat. The dog shows us a masculine body; but a feminine ether-body is woven into his nature, and it is this that sets the character of the whole dog family. These animals are humble, devoted to their masters; submissive and peaceable. The masculine physical body, however, makes them into warriors and protectors. In the wolf, these 'physical' qualities come more to the fore; likewise in the clever, often cunning fox. In the jackal, however, and especially in the domestic dog, the feminine etheric qualities make themselves felt more strongly.

Looking at the cats, we see they have developed a more feminine physical body. They are outwardly soft, fond of cleaning themselves, and display many other traits that we must call 'feminine.' Behind this, however, hides the masculine ether-body; it patterns the coat boldly, and gives rise to the wild, carnivorous disposition, the aggression and rapacity.

Rudolf Steiner adds the following characterisation:

The woman has masculine qualities within, the man feminine ones. So while the man becomes a warrior through his outer bodily nature – for this outward bravery is connected with the external organisation of his body – the woman possesses inner bravery, the capacity for devoted self-sacrifice.

These are precisely the etherically conditioned qualities of the dog; while the cat's bellicose behaviour breaks forth from her masculine ether-body.

'When a man becomes productive,' Rudolf Steiner continues, 'he goes out into what is around him. The woman acts in the world through devoted passivity.'

We might add, what the woman, through her physical nature, shows as passive devotion, appears in the dog family through their ether-body. In the same way, the physically based outward activity of the man becomes the cat's etherically-based readiness to spring into action.

Pursuing this line of observation a step further, we meet with a fundamental difference between human beings and carnivores. In the human race, sexual differentiation brings about a far-reaching separation into man and woman; in the realm of the dogs and cats it does not appear merely between males and females, but permeates entire groups of the animal kingdom. All dogs, whether male or female, almost uniformly show a male character. Similarly the cats, with the single exception of the lion, display qualities of the human female.

In human beings – and here alone – the sex difference penetrates so deeply into the bodily organisation and character of the individual that one could almost speak of two different biological species. The man is more masculine than any male animal; and the woman, in all her varieties, is more feminine than any female animal. For in the human race the division into man and woman occurred in such a way as wholly to pervade the physical-etheric body. The man has a masculine physical and a feminine etheric body, the woman a feminine physical and a masculine ether-body: by virtue of this fact the division was made complete. Two organic poles confront one another. Man and woman are an expression of two archetypal opposites.

This is never so with the animal. Although many birds display notable sex differences – in colour, markings, behaviour – the difference is minimal in comparison with human beings. Where it is pronounced – in the insect world, for example (one need only think of the drones and the queen bee, or the giant female termite; also the difference between male and female spiders) – it is no longer true sexual dimorphism, but a division of labour. The drones carry out the task of fertilisation, just as the queen has the job of laying eggs.

Having determined this, we can now begin to feel our way towards the polarity of dog and cat. In them, images of the masculine and feminine human bodies appear. Insofar as the physical-etheric nature of the human being is concerned, the dog is an image of man, the cat an image of woman. That is why these two animal groups occupy such a special position in their relation to the human being.

A folk legend from Palestine recounts:

> Once upon a time, when the world was young, to each
> and to every kind of animal a duty was assigned. The dog
> and the cat were relieved of menial duty because of the
> faithfulness of the one and the cleanliness of the other, and
> a written document was given them in attestation thereof.
> The dog took charge of it, and buried it where he stored
> his old bones. This privilege of exemption, however, so
> roused the envy of the horse, ass and ox, that they bribed
> the rat into burrowing down to destroy the charter. Since
> the loss of this document, the dog has been liable, on
> account of his carelessness, to be tied or chained up by his
> master, and the cat has never forgiven him.[12]

The story of the Fall – in disguise – sounds through this legend. Many other fairy-tales and legends dealing with cats and dogs point towards this mystery in a similar manner. One example is the story of the Town Musicians of Bremen, where the ass represents the physical body, dog and cat the etheric, and the chicken – crowing and beating its wings – is an image of the soul. It takes all four working together to act as a human being.

Cats, dogs and the threefold human organism

Looking back to the second section of this chapter and reviewing the morphological differences between dog and cat, we are now in a position to relate these to the differentiation of male and female.

We said that dogs have developed the limbs and senses better than cats. Cats in turn have developed the nerves and metabolic system much more strongly. We also found that dogs live more in the breathing, while cats are orientated more around the circulation of blood and the beat of the heart. Placing these differences side by side, we obtain the following summary:

Dogs	*Cats*
Sense organs	Nervous system
Lungs and breathing	Heart and circulation
Limbs	Metabolism

To those who have insight into the threefold structure of the human organism, which Rudolf Steiner repeatedly described from ever new angles from 1917 onwards, it will be immediately evident that this reflects in the above system. Here we see a double expression of this threefold principle, In dogs, the sensory system, the respiration and limb development are emphasised, revealing the 'masculine' character of this group. Are not men taller than women as a rule? He has longer, more strongly differentiated limbs and is orientated more around the breath. His chest is broader, his shoulders wider and his breathing deeper than hers.

The female organism – as with cats – is more strongly orientated around the metabolism. Connected with this are the monthly period, the ability to become pregnant and to supply the developing embryo with sufficient nourishment. Moreover, the entire circulatory system alters during pregnancy, the performance of the heart being considerably enhanced. Instead of the senses – which enable the male particularly to enter actively into what is going on around him – the woman acts in the world through 'passive devotion.' This she can do by virtue of her enhanced nervous system.

Now it becomes still clearer that the dog is determined physically by his masculine traits, but in character bears the stamp of his feminine etheric forces. The cat, by contrast, is physically an image of the feminine organism, while her behaviour reflects the masculine etheric forces. In the human being the determination is reversed; for it is our ether-bodies – feminine in the man, masculine in the woman – that mould the human physical constitution, while the physical body sets its stamp upon character. In this context we can only refer to these complex relationships in a cursory and generalised way.

Anyone who lives into these pictures and allows them to speak directly to his inborn power of judgment, will soon perceive the special situation of dogs and cats. They are expressions of a polarity which permeates all creation – that of the masculine and the feminine. Wherever and however they appear, they carry both aspects in themselves: the masculine dog shows feminine devotedness; the feminine cat takes masculine pleasure in attack and conquest. That is why Ishtar – the Venus of the Babylonian world – was invested with both feminine and masculine traits of character. As the evening star she appeared as the great goddess of love and devotion; as the morning star she revealed her youthful masculine power and might. In Assyrian portrayals she wears a man's beard, thus manifesting – like the god Ashur – her leonine nature.[13]

This now throws light on the nature of the 'king of beasts.' For, in all other cats, great and small, the masculine force is spent in their ferocity, but in the lions it penetrates right into their physical constitution. The mane comes into being to mark the regal lion – the morning star – as acknowledged monarch. Among all the cats, he alone receives a masculine body, despite belonging to a feminine race. In this he reveals his double nature.

212

This also came to expression on a different level in ancient Egyptian worship. Here there were two goddesses, both of whom bore a human form and a leonine head. Sekhmet the 'mighty' was revered in Memphis. 'She is the goddess of battles who, like the Uraeus Serpent of the kings, spits fire.'[14] At her side was the goddess Bast, who was soft and friendly. She bore the dancers' sistrum (rattle) in her hand, and a basket on her arm. The Egyptians experienced the two as 'one.' They would speak of a person who was as friendly as Bast and as fearsome as Sekhmet. In this way they referred to the mystery that Sekhmet represents the masculine ether-body, and Best the feminine physical nature of the cat.

The Egyptian gods with dog- and jackal-heads, on the other hand, were masculine. They were gods of the dead (Anubis), or gods whose images were led out in the forefront of battle. The Up-nats, the 'path-finders,' went forth before Osiris and were his comrades in arms. As their attribute, they carried club and bow. Thus the Up-nats pointed to the masculine physical nature of the dogs. Their heads were more like those of wolves.[15] Anubis, with his dog's head, served the deceased, whom he led to enlightenment after death. Here was an image designed to honour the faithfulness and power of devotion of the dog's etheric nature.

These mythological images reveal the open secret that dogs and cats bear with them: their masculine and feminine constitution, which has made them human companions since primeval times – though in their higher being, as group souls, they preceded him.

The genesis of dogs and cats

The genealogical theory as well as the paleontology of the carnivores are still wrapped in darkness. We know only that their earliest remains derive from the beginning of the Tertiary period, indicating that, like most mammals, the carnivores made their appearance relatively late in the earth's history.

From Rudolf Steiner we have quite exact indications about the origin of the carnivores:

> That which belongs to the lion-like animals did not begin to take effect on earth until Atlantean times, and it came to earth as though it were pressed out of the interior of the earth on to the surface.[16]

Thus we can picture the race of carnivores as originating at the beginning of old Atlantis. This great epoch corresponds approximately to the Tertiary period of geology.[17]

It was also at this time that the human physical body began to consolidate. Language began to take form, and the power of memory developed.[18] Presumably at the same time the first physical sex differences emerged in the human body; and what came to be a polarity then appeared – cast out of human evolution – as doglike and catlike animals.

It is not easy to imagine how those first carnivore-bodies were fashioned, but we can assume that they were still much closer to the human form than they were to be later on. Maybe the Egyptian statues of Anubis and Sekhmet are much closer to the truth than are modern paleontological conceptions of early animal forms.

In any case we must suppose that the different 'models' for man and woman, which arose as human bodily nature took form, became dog- and catlike beings. For tiger and panther, wolf and fox are trial variations – models – for the ultimate form of the male or female body.

This sexual differentiation developed quite late, as we learn from a further indication by Rudolf Steiner. He says that sexual reproduction did not enter in completely until the middle of the Atlantean period. Earlier, there was an intermediate stage. And of this intermediate stage we learn that there must have been a period of transition from sexual unity to division of the sexes; that there was indeed a certain state in between virginal reproduction where fertilisation came about through forces living in the earth and sexual reproduction.[19]

The transitionary period described here is the epoch in which we must seek the origin of the carnivores. It is the time of the first four sub-races of Atlantis, the epochs of the Rmoahals, Tlavatli, Toltecs and Turanians. And we can almost certainly place the emergence of the cat tribe in the first two epochs, and that of the dogs in the later two. Thus we can say that, half-way through Atlantean times, not only were man and woman fully formed in their bodily nature, but the ancestors of the lions, tigers, panthers, wolves, jackals and dogs also inhabited the earth.

Then, during the gradual collapse of the Atlantean continent, both groups of animals migrated eastwards with humankind. The imagery of Genesis expresses it by saying that they were taken along in Noah's ark.

During these migrations, however, the dog- and cat-tribes grew more distant from people and went wild. They became similar to present-day carnivores. Gradually they loosened their connection to humanity to seek their wild independence.

It was now the time when human beings had to find their own inner footing before receiving the strength to become the inaugurators of the post-Atlantean cultures on their path back from east to west. Upon their return, then, human beings reunited with their lost brothers – the dogs and cats. During the Persian culture, the second post-Atlantean culture, dogs were brought close to people; and in the Egyptian-Chaldean period the house cat came back into the human fold.

Of this also there is an indication in a lecture by Rudolf Steiner. He described how an ancient Persian teacher might have spoken to his pupils about the wolf:

> Think of the wolf. The animal living as the physical wolf you now see has fallen from its former estate, has become decadent. Formerly it did not manifest its bad qualities. But if good qualities germinate in you and you combine

them with your spiritual powers, you can tame this
animal; you can instil into it your own good qualities,
making the wolf into a docile dog who serves you![20]

Here the current assumption is confirmed, that we must look
on the wolf as the true forefather of the dogs, and that the period
of their domestication was in the ancient Persian culture.

The house cat, however, is known to derive not from the
European wildcat, but from the Nubian wildcat. The first
remains of the house cat come from Egyptian times. In the
prehistoric epochs of Europe – the Hallstatt and pile-dwelling
epochs, for example – no skeletal remains of cats are found. In
Egypt, however, cats were held sacred; they were embalmed and
buried in gigantic cat cemeteries, which each city had. This is
reported by Herodotus. It was not until quite late – in the first
Christian centuries – that the house cat found entry with the
Romans and Greeks, gradually pushing out the weasel and pole-
cat, which had been kept as catlike domestic animals.

In this way dogs and cats rejoined humankind. The prodigal
son who had been left by the wayside on the migration from
west to east – from Atlantis to eastern Asia – was taken up again
in Persia; while the daughter who had lost her way in the wilderness sought connection to humans again in Egypt. Since that
time, the two of them have accompanied us and become our
domestic animals in the literal sense of the word.

Once – when humankind was coming into being – dogs and
cats were cast out of humanity as trial models in the sexual differentiation of bodies. At that time they became witnesses to
the Fall. Today, after the Christ event has taken place, dog and
cat stand under the human 'I,' which has become active in the
human being. The dog becomes helpful to him; the cat seeks his
neighbourhood. Both live in the shadow of human existence.

Can we not see it as kind of prophecy that the constellation
of Aries, the Ram, which ruled during the epoch of the Mystery
of Golgotha, was seen with the likeness of a dog in China and

Indo-China, but as cat in Egypt? For Aries symbolises that chaste human nature in which there will be no more division of the sexes. It points towards future times when, under the Lamb, the sensual human form will be transfigured and reappear on a new level as an androgynous being. Then dog and cat will be united with humans again and will have achieved the redemption of their destiny.

11. Brother Horse

Introduction

The ideas generally accepted today about human and animal evolution are closely bound up with the image of a family tree. No matter what species or family of animals is under consideration, the moment we inquire about its origin, we are asked to think of a branch that long ago split off from a greater branch and entered specialised development. The larger branches of the supposed family tree are for the most part still great unknowns, which must be hypothetically assumed in order to remain true to the idea of the family tree. And as for the trunk of the tree, it disappears in a fog of vague conceptions

Modern paleontology and geology have left us with an endless multitude of 'branches' and 'smaller branches' which portray as their 'leaves' and 'blossoms,' the living and extinct species of animals. Many hundreds of thousands of different forms are now known and described. They can be gathered together into different classes, subclasses, and groups; this has given rise to the magnificent and fascinating system of zoology. However, as soon as one tries to trace these end-results back to more primitive forms – their original 'limbs' and 'branches' – one meets only with vague assumption and unproved hypotheses. I am convinced that a spiritually true ordering of the animals will be possible in the future only along the lines of Lorenz Oken's basic ideas. This great naturalist once formulated his view thus:

> It will be found in due course that the whole animal
> kingdom is nothing else than a manifestation of the
> single activities or organs of the human being; nothing
> other than the human being taken to pieces.[1]

If one really enters into this idea and begins to test its worth, it is soon found to be illuminating and helpful. What up till now had seemed senseless and accidental takes on sense and so becomes understandable.[2]

Instead of a genealogical tree, we see groups of forms, which can be related to larger or smaller groupings. The individual groups display similar tendencies. For example, the mammals of Australia – although they unfolded in completer isolation from the other mammals of the earth – have all developed similar forms. Richard Hertwig gives this phenomenon clear expression. He states:

> In their present area of distribution, the marsupials, in
> adaptation to similar living conditions, have gone through
> a development fully analogous to that of the placental
> mammals of the rest of the planet, so that complete
> parallel groups can be set up to the common mammalian
> orders (carnivores, rodents, insectivores, ungulates).[3]

In fact, living conditions in Australia were not similar to those in other parts of the earth, and also the marsupials did not have to 'adapt' to their environment. They were adapted to it from the beginning, developing forms, however, which became predatory, rodent, and hoofed marsupials. These groups are parallel developments to the three forms elsewhere. There are formative tendencies inherent in the mammals that bring out one of the three groups: the nerve-sense system (rodents), or the rhythmic system in particular (carnivores), or the metabolic-limb nature of the animal (hoofed mammals). Here, the threefold human being is the archetypal 'trunk' out of which the three groups of mammals unfold.

It would be very good to replace the word 'develop' with the word 'unfold' more often. This is much closer to the course of evolution, and would help to free us gradually from the narrow concepts brought in by Darwinism and Mendelism. There is no single genealogical tree along whose main axis all animals have developed. Each age of the earth created its own groups of forms; and in the midst of them the archetype 'human being' resides as the form-giving tendency.

All the great groups of forms – such as the fish, arthropods, echinoderms, mammals, and so on – have a number of smaller groupings associated with them that bear within their range an image of their archetype. That is why the marsupials of Australia unfold in a way similar to the mammals of the other continents. They form hoofed, rodent, and carnivorous groups, and each of these shows similar unfolding tendencies.

Thus, within the order of the carnivores an entire group has committed itself to life in the water: the seals. A similar correspondence is found to the order of the hoofed mammals (ungulates). Here it is the dolphins and whales that have become aquatic mammals. In the group of the rodents, the beaver shows the aquatic tendency.

These ways of unfolding have nothing to do with adaptation. They reveal simply that every group of forms represents a self-contained unity obeying similar formative tendencies. Seals, dolphins and whales, beaver: they are all aquatic members of their higher order.

Even the birds send a part of their living totality into oceanic regions: in the Antarctic, we find penguins, and in the Arctic the Great Auk survived into the nineteenth century. Out of completely different families similar forms arose at opposite ends of the earth.

The whole kingdom of animals, in all its forms, is an ever-repeating example of this will-to-unfold. Behind it lies an all-encompassing law: and this law is the human being himself. He is not the crown, but the inmost kernel, of all creation.

In every lesser or greater formal group, this kernel manifests in such a way that a particular species tries to approach perfection by coming as close as possible to the form that this group seeks to express. All other species and families will find their place as variations around this central figure. Through an intuition of this, that was still influential in the past, people called such perfected forms 'kings'. They called the lion the king of the animals, the eagle the king of the bird world. Even today, the especially large and colourful penguins are called emperor penguins. And who would dispute that among the dog-like animals the wolf comes closest to the archetype striving for realisation there?

Of course, we can no longer hold to concepts so tinged with feeling; too many sympathies and antipathies would cloud a true picture of the phenomenon. Nevertheless, we must begin to learn to think in new categories if we are to come closer to the form-giving tendencies active in nature.

The hoofed or pasture animals

The horse is a particular member of the great group of hoofed animals known as ungulates. These in turn belong to the vastly greater order of placental mammals, that is, those animals which develop in the maternal organism by means of a placenta.

The hoofed animals themselves – following Richard Owen's brilliant insight – have been divided into two sub-orders. One includes those with an odd number of toes on each foot (the perissodactyls), the other those with an even number (the artiodactyls).[4] The odd-toed are the much smaller and – in their species and families – more limited group. Besides the horse, it embraces the tapir and rhinoceros. The horse family itself includes the ass and zebra.

The second sub-order, the even-toed ungulates, is incomparably richer. Here, two main groups are distinguished: the ruminants and the non-ruminants. The latter includes the hippopotamus and the large family of the swine. The ruminants embrace the groups so close and well-known to us, the creatures that have formed our picture of the animal kingdom from childhood onwards. Here are the cows and camels; the llama and the giraffe; bison and buffalo; sheep, goats and ibex. Here are the swift gazelles and antelopes; the deer, stags, reindeer and elk, and countless others. With many of them our being is intimately bound up in childhood; they live as close companions in our soul's inner space.

All hoofed mammals, both even- and odd-toed, have many common features. None of them is excessively large of strikingly small. Rather, the kind of variation they display ranges between clumsy and graceful. From the unshapely rhinoceros and hippopotamus, it passes through bison, cow and camel all the way to stag, deer, antelope and gazelle. This scale gives a vivid picture of the variability possible in hoofed animals, as determined by body covering and musculature. The body of the rhinoceros, covered with great plates of horn that envelop him like a coat of armour and thicken his legs into columns to support this mass – that is one pole. The other is represented by the graceful, delicate gazelles and antelopes, whose slender, nimble limbs carry the dainty body with a light spring over steppe and prairie.

Only giraffes stretch their strong necks up to the treetops, breaking the rule of size among the pasture animals. Most other genera and species keep to the human measure; it is hardly ever surpassed. Moreover, even the tall giraffe has a comparatively small body; only neck and limbs are pulled out beyond this mean. Clumsiness and grace, unwieldiness and elegance are the only factors at work here.

Another essential feature is perhaps still more striking, since it expresses itself in new forms of organic structure: on the ends of the feet, hooves form; while on the head a great variety of bone and horn appendages appear. It is one of the strangest organ-forming capacities of the pasture animals that the males, chiefly, grow antlers and horns, while the limbs are closed off and hardened with hooves.

We must learn to recognise these two formations as belonging to one another morphologically; upwards, the process on the head; downwards, the hooves. To look at them according to their use leads us on false paths; for horns and antlers cannot be seen as weapons in the battle for existence, nor are the hooves a help for swifter mobility. If one animal wounds another with a kick of the hoof, or attacks an enemy with horn or antler, this behaviour recalls the famous anecdote, which tells how Luther hurled his inkwell at the devil. As little as an inkwell can be called a weapon, so little are horns and antlers instruments of attack or defence.

Both are very far from being any sort of tool. They are formations sprung either from the skin (horns) or the skeleton (antlers), transmitting upwards and outwards that which could not unfold completely within and below. The feet, closed round with hooves, are sealed off, as it were. Upwards, however, a new organ unfolds in wonderful variety and beauty.

Hooves and head processes are forms that determine one another; this becomes evident when we simply compare even-toed and odd-toed animals with each other, for the even-toed ungulates also carry horns and antlers arranged in pairs. These develop as a bilaterally symmetrical organ. The rhinoceros, by contrast, as an odd-toed ungulate, bears its appendage in the middle of its face. Even if there are two horns, they are set in the midline of the nose, not next to one another. This shows that 'odd-toedness' is effective also in horn formation, just as 'even-toedness' manifests itself in symmetrical horns and antlers.

We have now found evidence of another basic principle that determines the forms of the hoofed mammals. Clumsiness or grace determine their body form; this form itself, however, exists in a field of tension between head appendages and hooves. Two formative tendencies interact here: one that moulds the outer form of the body, and one that seals it off below while opening it above.

The force determining the outer shape of the animal is intimately connected with hair formation, and hence also with the production and storage of warmth. All hoofed mammals are warm-blooded. The force moulding the bodily form has its root in the blood and hair.

The tendency giving rise to hooves, horns and antlers derives from the skin and the bones. Calcium and silica are at work here, shaping all the various forms of the head processes.

Horn gives the head a certain measure of weight – we need only think of buffalo and bison, the bull gnu and bongo. Only when the mass of the horn is released into strong winding (kudo, antelopes, chamois, ibex), does this loosened form come closer to that of antlers. These latter draw the head up and out beyond itself. Like a sense-organ, feeling and sensing, they extend into the world around. Calcium-forming forces are at work in horn, while silica forces determine the form of antlers.[5]

What remains hoof-like in horns is overcome in antlers. Poppelbaum characterises this polarity thus:

> In the horn, the skull pushes outwards into the world a little
> way; through the dying of its casing, however, the horn
> closes itself off and stops up the formative forces of the head
> ... The antlered animals, on the other hand, continually
> open their head to the outside. Hence their temperament
> is also more lively, more awake and nervier; their eyes have
> more expression and their movements more grace.[6]

Only one group seems to escape the general rule which we have described: the horses. They bear the most strongly and perfectly formed hooves; yet no outgrowth crowns their forehead.

It could be objected that the swine, the hippopotamus and the tapir also lack horns. Instead, however, they have each developed some sort of snout which makes up in nose-development for what otherwise grows on the frontal bones as horn or antlers. Moreover, swine and hippopotamus often develop great tusks, which fulfil in the jaw what is absent from the head.

The horse – along with its closest relatives, the ass and zebra – raised its head out above the body, sensing and scenting, its large nostrils opened, not burdened or lightened by horns or antlers. This gives the horse a special position among the ungulates.

The horse and the ungulates

We have already mentioned that the group of the odd-toed ungulates is small in comparison with that of the even-toed. Nevertheless, this order includes a great number of species which are today extinct.

> The history of the earth reveals ... that they [the odd-toed ungulates] have already passed their prime. Trouessart counts 131 fossil genera and sub-genera with 517 species and sub-species, and only eight living genera and sub-genera, with 36 species and sub-species.[7]

The even-toed ungulates, on the other hand, are still growing and unfolding as a group, with a large number of genera and species inhabiting all continents – Australia and the Antarctic excepted.

Thus the horse seems to project out of past ages of the earth into the present, and yet has become one of man's closest companions. Or has it been his brother from the beginning?

There are few genera of animals over whose paleontological development we have so clear a view as we do of the horse's. Its early phylogeny can be gathered from remains found almost exclusively in America. The earliest form stems from the Eocene, the first epoch of the Tertiary period, which we may equate with the early Atlantean development. Here the horse occurs as eohippus *(Hyracotherium)*, and during this time unfolds into orohippus and epihippus. These three ancestors of the horse show a clear similarity in build to the present-day horse, but they were very much smaller. Eohippus and orohippus were about the size of a domestic cat. Also, the development of the foot had not yet gone in favour of the middle toe; the second and fourth

toes were still clearly present. And the teeth were not as specialised as they are in today's horse.

We can picture the first horse-like creature as small and nimble. Abel, an expert in this field, writes:

> The oldest horses were no steppe-dwellers; they were
> little animals, and in their general appearance must
> have come much closer to a Chilean pudu or a Javanese
> kanchil (*tragulus*) than to a small modern horse.

These two – the kanchil and pudu – are among the smallest
living hoofed mammals. They grow no higher than about 20–30
cm (8–12 in), and their length, including the tail, comes to 40–50
cm (16–20 in). The kanchil is a chevrotain or a dwarf musk-deer,
the pudu a true deer. Abel continues: 'Matthew has pointed out
that the oldest horses probably inhabited the thickets ... living
chiefly on soft leaf and juicy herbs.'[8]

Then their unfolding proceeds. Through the Oligocene,
Miocene, and Pliocene – throughout the epochs of the Tertiary
period – the horse ancestors grow in size. The mesohippus,
merychippus, and pliohippus take form, the last-named almost
reaching the size of our present equides. At the same time, the
lateral toes dwindle and disappear; and the middle toes alone, on
front and hind feet, bear the growing weight of the new powerful
body. The dentition also becomes unified; the eye-teeth atro-
phy, and the premolars are transformed into molars. The figure
clearly shows the change in size and form of the head in certain
Atlantean species of horses.

Many more details could still be revealed from the numerous findings; we will not go into them all. One fact, however, is
significant: as soon as the type pliohippus is attained, it dies out
on the American continent. In the Pleistocene deposits, horses
are no longer found 'until the day when they were introduced
again by man.'[9]

Zoologists and paleontologists are astonished at the sudden

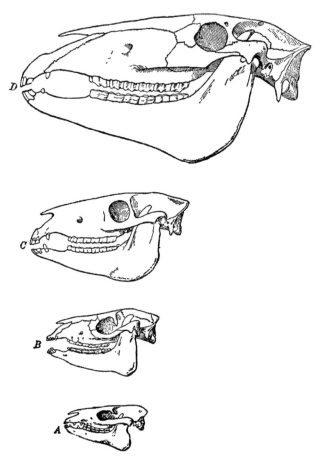

1. Skulls of four horse ancestor, in equal reduction. A: echippus; B: meschippus; C. protohippus; D. equus (recent) (From Abel *Paläobiologie und Stammesgeschichte,* Fig. 148)

disappearance of the early horses living in America. Some suppose that an epidemic nearly exterminated the race of horses there, and that only small groups escaped by way of Alaska and the Aleutian land bridge to Kamchatka and eastern Asia.

In fact, it was not until the sixteenth and seventeenth centuries, after the rediscovery of North and South America, that the Europeans brought horses there again. None of the Native

Americans or Aztecs living at that time had known these animals before. And yet the original home of the horses and hoofed animals was the American continent. As soon as one introduces findings from Rudolf Steiner's spiritual research about old Atlantis, one finds a key to the evolution of the equides; and the riddles begin to be solved. This great continent collapsed in vast floods at the end of the Ice Age and piece by piece Atlantis sank into the widening Atlantic Ocean.

But in Europe and Asia the escaped remnants unfolded – under quite new conditions – into the horses of today. It is now that they become animals of the wide steppes and grasslands.

About their spread, Abel writes:

> Today horses live in the cold, dry highlands of central Asia, as well as in the Masai steppe of Africa, where the zebras ... form enormous herds, or did so until recently. In the European ice age, however, we have found not only horses corresponding to the Asiatic horses of the high steppe – thus proving the occurrence of Pleistocene steppe horses in Europe – but also horses that lived in the forest or the tundra, as we can infer from the geological conditions and the character of the accompanying fauna.[10]

Though we have given only a brief indication of these phenomena, they permit us to make a preliminary picture of the unfolding of the horse. Out of small, almost insignificant mammals, dwelling in the thick bushy undergrowth of the Atlantean world, larger and larger animals developed, who by the end of the Atlantean period had reached the form and stature of modern horses.

At the same time all other mammals unfolded, especially the hoofed and pasture animals. Alongside the horse arose tapirs and swine; giraffes and rhinoceros took form; buffalo and bison, sheep and goats, antelopes and red deer developed. It is the non-ruminants that go back farthest into the early Atlantean period:

swine, hippopotamus, and the small race of the babirusa. Later, in the Oligocene, came the Tylopoda such as the llama and the camels and dromedaries. At the same time appeared the dwarf musk deer (we have already mentioned the tiny kanchil). Then, from the Miocene period on – corresponding to about the middle of the Atlantean period – all the ruminants enter into the development.[11]

We can see here such an endless abundance of forms, shapes and kinds. All are hoofed mammals; but they are also horn and antler bearers, and they have one characteristic – not yet mentioned – which deeply affects their temperament and being: they are all chiefly plant-eaters, having gradually become creatures of the woods and pasture. The predators which arose at the same time are true hunters, while the hoofed mammals grazed the steppes and prairies. They lightened the dense Atlantean forests and overcame the luxuriant nature. Through their mode of living, the Niflheim of the world begins to be brightened with the light of the sun. The newly arising presence of the hoofed mammals cleanses the thick atmosphere, and clears the haze and mist of the Atlantean realms; it brings digestive, catabolic forces to bear against the wildly proliferating plant growth.

Here we meet one of the mysteries with which the ungulates confront us. They – the ruminants above all – transform green plant matter into white milk. This is a metamorphosis through which light enters the earth-world. Once – in primordial times – the whole atmosphere was permeated with cosmic milk. At that time, breathing and the taking in of nourishment were still one process. Later, this unity divided, and air and food consumption become two independent processes.*

It was the ungulates who activated this development. They

---

* See Rudolf Steiner's description: 'When the Lemurian evolution was in its prime, neither the breathing system nor the alimentation system existed in the forms we know today. Substances were then quite different; breathing and nutrition were in a certain sense connected; they performed a common function which was not divided until later on. The human being absorbed a kind of water, milky substance and this supplied him with what breathing and eating give him today.'[12]

not only roamed the land; they cleared meadows and woods, transforming the cleared green into nourishing white. They were preparing for those who later become farmers. And so it is no longer strange that cows and sheep, goats and swine are found with people again; for it was they who first prepared the soil where it could later be ploughed and harrowed, sown and harvested.

And where do the horses fit into this story? Do they pass through it untouched? Are they not among the oldest of all ungulates, reaching back to the very beginnings of Atlantis? They took on neither horns nor antlers, but held to the middle road of unfolding so that the others could deviate in other directions and fulfil their specialised tasks.

Throughout the period of Atlantis the horse developed as a king. Its intention was to set only the middle toe of hand and foot against the earth, so as to attain a perfect balance between gravity and levity.

Just as the human being is the measure of all things, so the horse can be called the measure of all ungulates. Single-mindedly it unfolds towards its earthly task: to carry humans on its back, to draw their vehicles over roads and tracks. The horse develops towards humankind: all ungulates serve the earth. In this way the horse becomes our brother, and the hoofed mammals our kin, closer or more distant.

Hooves and limbs

In no other group of ungulates have the hooves reached such perfection as in the equides. Horse, zebra, and ass have developed these terminal organs to such purposeful completion that one is scarcely reminded of their origin.

Eocene		Rhinoceros	Eohippus ↓ Orohippus			Tapir	
Oligocene	Babirus Hippopotamus		Mesohippus ↓ Miohippus		Swine	Dwarf musk de[er] Llama Camel	SL..
Miocene	PLUMP FORMS	Buffalo, bison European bison Aurochs ↓ Cattle	↓ Parahippus Merychippus ↓			Goat, sheep ibex, chamois Gazelle Deer, stag	FC Giraf[fe]
Pliocene		Antelopes	Pliochippus Hipparion ↘ Ass ↘ Equus ↗	Hippidion Zebra			
Piestocene		↓ ↓ ↓ ↓	↓			↓ ↓ ↓ ↓	↓

All hooves have developed out of the rudiments of fingernails and toenails; and in most even-toed ungulates this is still obvious, because the horny hoof portion usually covers only the front end of the foot. The horse's hoof, by contrast, forms an almost closed ring round the large middle toe. The anatomical structure of this organ is complex; it is such as to permit simultaneous firmness, elasticity, and continual renewal. The hoof does not consist of a simple flattened plate, but of many small horn tubes which, closely joined to one another, provide stability and elasticity. In the centre of the hoof lies the final phalanx, the 'coffinbone,' surrounded and cushioned by the 'frog,' a mass formed of cartilage, fat, and connective tissue. In the ring of the keratinised hoof-plate, the frog forms the elastic layer which receives the pressure of the coffin-bone and transmits it to the inner wall of the hoof plate. On the underside of the hoof, the frog is covered and protected from direct contact with the ground by the horn and leather covering of the sole.

The hoof is a kind of joint. The toe bone is the head, the hoof proper the socket, and the frog the bearing. All joints, however, are extremely sensitive organs. They should really be described as sense-organs in which the senses of touch, life, and self-motion work together to provide a continuous, dim awareness of the position and posture of the body. Hooves are subliminal sense-organs which permit the horse to sense and feel out the ground surface in such a way that each hoof-beat becomes a sensitive experience. How sensitive horses are to any hoof disease or to a badly sitting horseshoe! There was a time when any farrier could speak of that. Thus the hoof is not only an organ of protection, as is usually assumed; it is much more a sense-organ, and could be compared to a delicately built, enlarged feeling-eye or ear. In the horse this organ seems to be especially well developed.

When one considers this fact, the question immediately arises: What gave rise to this development? Why is it that hoof organs were formed on all ungulates, but in a special way in the equides? Can it be understood from what has been presented so far? Indeed, I venture to ask: Why do the ungulates have hooves?

Let us go back once more over the evolution of the horses; let us look at the gradual development of the limbs, which can be followed in the figure overleaf. Here the successive stages in development of the front foot are represented from mesohippus to pliphippus. What do they show? First of all we can recognise the lengthening and enlargement of the middle toe-bone. From type to type, the proportions of this bone increase. However, at the same time one can observe the toe turning up away from the ground, stretching more and more. The angle between the toe bone (phalanx) and mid-foot bone (metatarsal) opens appreciably, and continues to grow. This means that the foot is moving from a more horizontal to an increasingly vertical position. Here is clear evidence of the process of becoming upright which must once have taken place in the horse's foot. One can see how the fourth toe is gradually lifted further off the ground, carried up

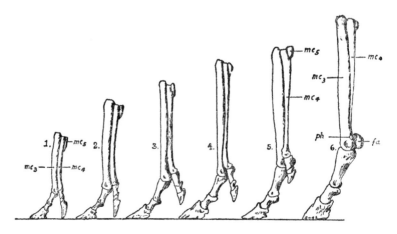

Side view of the bones of the left front foot of six North American ancestors of the horse from the Tertiary (to scale, after Abel, *Paläobiologie und
Stammesgeschichte,* Fig. 78).
1. *Mesohippus* 2. *Miohippus* 3. *Parahippus* 4. *Merychippus primus* 5. *Merychippus
eohipparion* 6. *Pliohippus*
mc 3, 4 and 5 are third, fourth and fifth middle digit bones

by the middle toe as the latter stretches out; and since it loses the
ground beneath it, it degenerates.

Rudolf Steiner often spoke of the force of uprightness which
passes through the human being and gives him his vertical posture. The same force – to a lesser extent and in an altered way
– also works in the animal kingdom. Among the ungulates, and
in particular the horses, it can be detected in the evolution of the
limbs. What can we conclude from this?

During the Atlantean epoch, all hoofed mammals went
through a process of becoming upright – not like the process
which penetrated and formed the human being, but similar to it.
The way the force of uprightness worked in the ungulates was in
stretching their legs, standing them on their toes, and thus lifting
them a little against the earth's field of gravity.

Little children, when asked how big they are, will stand on
tiptoe and hold up their arms; this is what the hoofed mammals
once did. They pushed up a little against the earth's field of
gravity by stretching their limbs. Some animals even wanted to

imitate man and tried to erect their spines; but it remained – as in the bear and kangaroo – a rather pitiful attempt.

Thus it was no sort of adaptation that led to the special development of the single toe in the horse. And why should a creature fare better in the 'battle for existence' on one toe than on five? After all, the even-toed ungulates have survived the last millennia quite well, although they have chosen to step and jump on two toes instead of on one. And four- and five-toed animals move just as quickly as one- and two-toed. There never was a process of adaptation involved here!

Rather it was the process of becoming upright, of straightening up, which ran through the ungulates. The horses had the courage and endurance to stand on the middle toe and finger alone and to stay poised in that position. The even-toed ungulates preferred to choose a somewhat securer equilibrium, balancing on the third and fourth toes.

Both the single-toed and the even-toed ungulates acquired hooves, for something had to counteract the erecting force which penetrated the limbs. This powerful upward-striving tendency was given a lower termination: the increased horn formation created the hoof. It is a unique acquisition in the realm of animal forms.

This stretching and erecting works not only in the limbs; the same process penetrates the whole animal, and at the upper pole – the frontal and nasal area – it leads to the breaking through of horns and antlers. They are like the raised arms of the little child stretching up. Because the ungulates stand up on their toes and alter their response to gravity, they move more into the field of levity surrounding the earth. They 'scent' this new sphere; the animals, poised on their toes, are now opened to a light-space. This light-ether domain likewise forms its own organs: horns and antlers.

Only the horses – along with ass and zebra – remain free of these organs. Their hooves have come to such perfect development that they have used up the formative forces down in the field of gravity, leaving none for any upward development.

However, by no means do all horned animals raise their head up into that brighter sphere: gnu and bison, buffalo and many sheep tend to hold their horns down towards the earth. It is only the antlered creatures which take this lifting and illuminating process to perfect fulfilment. The counter-force of gravity, one could call it 'levity', has taken hold of their skeleton, bringing forth a rising formation. Every year the antlers are cast off and renewed, each time appearing in a more splendid and powerful configuration. Stag and wapiti, roe, elk and reindeer are the crown of the race of hoofed animals.

Something which at first presented itself merely as a phenomenon now begins to disclose its secret – really an open secret.

The force of uprightness acts both upwards and downwards in the ungulates. At the pole of gravity it gives rise to the hooves; on the opposite pole of levity it becomes the creator of horn and antler.

Both of these – like the hoof – are disguised sense-organs; they expand the sphere of sensitivity of their bearers. The horn listens more to the body; it dimly takes in the rushing of the blood and the murmuring of the vital fluids within. The antlers, probing out into the world, perceive the air-currents which carry smells and colours to the animal.

This polarity reveals itself also in the substances of which horns and antlers are made; for bone tissue is different from skin and horn. The horn always strives towards rounding; the antler forms axial structures. The antler manifests the outward-striving force; the horn shows the closing off tendency which is produced by tangential forces. The antler grows from the centre to the periphery; the horn, by contrast, is formed concentrically out of the periphery towards the centre. The antler bears the radiating force of sight in it; the horn, the stillness of listening. And the hoof?

Rudolf Steiner showed in his lectures how in the structure and organisation of the three ossicles of the middle ear we can recognise a metamorphosis of the arm and leg bones. Hammer, anvil, and stirrup, which are stretched between the eardrum and the inner ear, represent the three transformations of the foot and tibia, knee-cap and femur.

> Just as human beings feel the ground with their two legs, so they feel the eardrum with the foot of the small ossicle. But their earth-foot, with which they move about, is crudely constructed. With the soles of their feet they coarsely feel the ground, while with this (transformed) hand or foot within the ear they continually sense the subtle trembling of the eardrum.[13]

In a similar way all ungulates, but especially the horses, sense the vibration of the earth under their feet. Because they touch the earth with only one or two toes, their stepping becomes much more of a sensory process than a mechanical action. Whoever has seen gazelles and stags, roe deer and chamois jump and climb, cannot escape the impression that they float or glide. A chamois hardly touches the ridge it is climbing; and an antelope flies over the wide expanses of meadow as if it were carried by the force of levity. The hoof listens; it senses the vibrations of the earth trembling under its light beat and gives the animal the feeling of certainly in overcoming gravity.

In watching horses when they are chasing, running, galloping, one can stand there amazed at this power-flooded grace. And one gets the impression that the rhythm of the hoof-beat, accompanied by its stamping melody, is a necessary component of their unique movement ability. The horse's hoof 'hears' the rhythmic sound of the impact, and in this way gives the beat for the flow of movements. Like a conductor, the hearing hooves direct the motor sequence.

The antler-bearing animals open up the etheric realm of light. The horned ungulates have a dull awareness of the waves of sounds within their own body. The horses, by contrast, create their own sound-world as they step and jump; and this is dully perceived by the hoof. All three are sense-organs: horn, hoof, and antler.

The gait of horses

Although the horse is among the largest of hoofed mammals, with powerful legs and a massive body, its mobility is astonishingly varied. Graceful, fiery, delicate, angry, powerful, slow, relaxed: these are just a few of the many registers on which it can play. The different possibilities of movement which have become specialised in other ungulates – we need only think of the gazelle, cow, ibex, giraffe, and bison – are united in the horse. Hence it is also the only animal that learns to move to the beat of music and can be trained to use forms of movement and gaits which are not natural to it. The Spanish Riding School of the Lipizzaner horses in Vienna attains the peak in this. To watch these riding arts can be a thoroughly musical experience.

Through the variety of its movement capabilities the horse confirms its central placement among the ungulates. It unites the possibilities of movement which we find divided up among the other species.

Although massive and heavy, the horse's body is imbued with clear harmony. Despite the thinness of the legs as they taper downwards, despite the contrasting size of the hind thighs and the body hanging between the four leg-posts, despite the excessively long head and sturdy neck, grace and dignity are always present to some extent. The beauty and harmony which meet us have less to do with the form of the body than with the grace of its motion.

And then there is the inner mobility and sensitivity of the equine nature. A slight trembling passes continually through the body; arteries appear under the skin and disappear again. The skin itself forms fine folds and wrinkles, and then relaxes once more. It is as though waves of sensation flowed ceaselessly through the body.

Is not the horse's body comparable to a musical instrument on which movement plays its rhythms and melodies? Seen from above, is it not like the body of a great cello? The head and neck of the horse would correspond to the neck of the instrument, and the characteristic doubled shape of the sound-box, with projecting curves in front and back, would be the outline of the torso.

Just where the bridge would be is where the saddle sits, and precisely beneath this point is the centre of gravity of the horse's body. It is located about twice as far from the back as from the belly, within the body, in the region of the heart. Instead of the strings there are legs, which make the movement possible. Movement plays on the instrument of the horse's body; the body vibrates in its rhythm, and the hooves are the dull sense-organs for the arising sensory-motoric vibrations.

It is only from this point of view that we can begin to understand the three or four natural gaits of horses.[14] Depending on personal preference, the amble may or may not be reckoned among the innate gaits. To begin with we will disregard it, and consider the three generally used gaits: *Walk*, *trot* and *gallop* (the canter is a form of gallop).

The characteristic element of the *walk* is that no two legs are moved at the same time. As in all gaits, the initiative proceeds from the hind quarter on one side; then the opposite front leg. It is a quick succession of crosswise movements, which permits a steady progress forward. The walk is never a kind of run, as are the other two gaits.

The *trot* has a different character. Although there are variants of this gait, its basic form is always the same: diagonal leg pairs move simultaneously, the right hind leg and the left front leg. The quick one-by-one succession of the walk is linked here into a two-by-two; only two hoof-beats are audible in a unit of this gait.

The *gallop* is a remarkably complicated movement. Here again, one of the hind quarters begins, but it is followed by the pair of legs lying diagonally to it, after which the remaining front leg finishes. If we picture the horse to ourselves as floating in the air, we can say that it is the hind leg which touches the ground first; then comes the diagonal leg pair, and finally the remaining front leg.

Since the *pace* is not natural to all horses, we will describe it briefly for the sake of completeness. Here the limbs of one side move together, alternating with those of the other side. Thus, the right side will move forward once, then the left side, etc.

If we try to make a picture of the three gaits, it might look as follows (1 and 2 indicate the number of legs placed down at the same time):

Walk:	1 – 1 – 1 – 1	1 – 1 – 1 – 1	1 – 1 – 1 – 1
Trot:	2 – 2	2 – 2	2 – 2
Gallop:	1 – 2 – 1	1 – 2 – 1	1 – 2 – 1

This makes it clearly visible that these three gaits – the amble is only another form of trot – are three different kinds of rhythm, based on the beat of the particular gait. The walk proceeds in 4/4 time; the gallop in 3/4 time, and the trot in 2/4 time. What could show the musical nature of the horse's movement more vividly?

Here the same rhythms appear which are basic to all folk dances. The waltz, two-step, csárdás, and many similar dances are built on these rhythms. They are inscribed into the movement configuration of the horse as inborn rhythms; and thus these animals become the natural bearers or rhythm and beat. The four extremities bring them to expression, and the horse-nature manifests itself in them. Walk, trot and gallop are the three archetypal rhythms of all music; the horse is their body. Moreover, this body is the image of the stringed instruments – which originally brought only rhythm and beat to expression.

It is only with prolonged and constant training that the horse is able to master artificial dance steps. Only the three rhythmic gaits described are natural to it; and here we immediately see the limited range of its motor possibilities, despite their relative variety. For the horse, only forward motion comes easily; sideways movement is hardly present. Nor is there a great degree of movement upwards and downwards. Thus the existing beat and rhythm lack any melody; this can be given only through training by a human being.

The spine lies horizontally; although the legs have taken in the force of uprightness, it has not grasped the vertebral column. For this reason the main plane of symmetry in the horse's body is sagittal, dividing the body along the spine into two symmetrical halves. The four gaits are centred around it, alternating from one side to the other; and the centre of gravity is also found within it.

The frontal plane can be conceived only as going through the front legs, perpendicular to sagittal plane. This is the plane which is pushed forward during motion, and in which up-and-down movement also takes place.

The horizontal plane, on the other hand, is the ground which the hooves meet. It is cut perpendicularly by the frontal and sagittal planes, which move forward along it. The bodily instrument of the horse, and the rhythmic motor activity which plays upon it, are inscribed in these three planes.

If any melody is to sound through this rhythmic locomotion, a rider must mount the horse. As soon as this happens, gait and direction – under the rider's control – can become melodically effective. Now the frontal plane passes through the human being, joining rider and horse together; the two have become a unit. The horse gives the rhythm, the rider the melody, and together they become a harmonious musical entity. This is one of the numerous secrets of the joy in riding: human being and animal together become music – music that can be experienced.

The horse has not found fulfilment until a person mounts it, rides it, and is carried by it. Though horses can be very beautiful in freedom, without humans they seem incomplete and naked.

Only through humankind do they receive their fulfilment, and the completeness due to them.

Out of small, inconspicuous creatures our horses have unfolded. They stayed in human proximity; but it was not until late that they became our brothers. Then they remained united with us over thousands of years, until modern technology has estranged them from our working lives again. Their own proper rhythm has been replaced by two- and four-stroke engines, and in developed countries their serving power is seldom in evidence.

Human being and horse

The close community of life which has formed between human beings and horses in the course of cultural development finds eloquent expression in different languages. In German, there is probably no other animal with so many different names. German speaks not only of a *Pferd*, but also of *Gaul* and *Ross*. The region where *Gaul* is used lies mainly in central Germany, surrounded in the south by *Ross*, in west and north by *Pferd*.[15] The word *Gaul* comes from the Middle High German and once meant a male animal. *Pferd* is derived from the late Latin *veredus*, a post-horse. And *Ross* – Old High German *ros*, Anglo-Saxon *hors*, and English *horse* – points back to a Germanic form originally meaning 'to jump.'

If one tries to feel one's way towards the inward significance of these three names, *Gaul* would probably designate more the draught-animal, *Ross* the riding animal, and *Pferd* the leader. *Ross* and *Reiter* ('rider') are closely related terms; *Gaul*, on the other hand, describes the rounded effort of an animal harnessed to a wagon; while *Pferd* has an element of pointing the way.

In English too we find a great variety of names. There are the sexual designations *mare*, *stallion*, and *gelding*; and the young *foal* may be either a *filly* or a *colt*. There is also the *stud* or *stud-farm* where horses are bred

By colour, horses are known as *chestnut* or *sorrel, bay, buck-skin, roan* or *dun.* We also speak of an old *nag,* a *jade,* or a *hack.* According to their breeds, we call them *thoroughbreds,* and *half-breeds, cold-* and *warm-bloods* and, of course, *ponies.*

Some famous horses had names which are still known to us: for example, Don Quixote's Rocinante and Alexander the Great's Bucephalus. Caligula's horse Incitatus was given the title of Consul by his master, and was honoured, kept and fed accordingly. These are clear signs of the brotherhood that grew up between man and horse.

It was an unquestioned custom among the Germanic tribes to bury warriors along with their horses. Attila too, was laid to rest together with his horse. In Denmark, in ancient times, 'during the construction of a church a live horse was buried in the foundations.'[16]

However, to suppose that horses have accompanied human-kind since primeval times would be wrong, as historical and pre-historic evidence shows. It was not until relatively late – around 3000 BC – that horses became domestic and work animals. Long before this people had hunted them. Many Stone Age depictions found in the caves of Spain, southern France, and Central Europe portray wild horses alongside aurochs, mammoth, elephant and ibex.

During various periods of the Ice Age, the wild horse, along with the wild buffalo, formed one of the most important sources of human food. At some Pleistocene settlements, its bones are piled in great refuse heaps.[17]

The wild horses of that time possessed a larger head, stronger teeth and more powerful jaws than the horses living today. Judging by the bone remains, they must have inhabited south-ern Europe in enormous herds. They gradually disappeared from there, withdrawing to the forests of the north. There they remained, living wild, right into the Middle Ages; and it is only in the course of the eighteenth and nineteenth centuries that they have died out. Pliny the Elder writes: 'In the north one finds herds of wild horses.' Centuries later – in the year 732 – Pope Gregory III writes to Boniface: 'You have allowed some people to eat the flesh of wild horses, and most people also that of the tame. Henceforth, holy brother, permit this on no account.'

Helisaeus Rösslin (whose name means 'little horse'), as late as 1593, writes of the wild horses in the woods of the Vosges: 'The wild horses are much wilder and more skittish by nature than the deer in many lands, and also much harder and more toilsome to capture.' He is of the opinion, however, that once caught, they are 'the best horses, equal to the Spanish and Turkish.'

Even today there are probably wild horses, the Tarpans – living in the steppes and forests of eastern Russia.[18] They are small animals, with thin, powerful legs and a long neck. They are similar in appearance to the ass, like the Przewalski's horses first discovered in 1879 in Central Asia. The latter live in small herds, each led by a stallion.

Perhaps it is there, in the land of the Kirghiz and Tartars, that we may look for the home of all post-Atlantean horses and their descendants. This is only an assumption; but there is little doubt that the Mongols of North and Central Asia used the horse for riding in very early times. Only much later did it reach the civilised nations. In Pritzwald's account:

> The horse did not number among the domesticated
> animals of the Sumerians and Old Kingdom Babylonians;
> for Hammurabi does not list it in his Code among the
> domestic animals. It is also absent in Egypt up to the
> Hyksos interregnum, as well as in the earliest history of
> the Israelites: among all the domestic animals enumerated
> in the possession of the patriarch Jacob, no horses are
> mentioned.[19]

Another description tells us:

In the Old Kingdom of Egypt, the horse was altogether unknown; the ass was the only beast of burden and work which was kept ... It was not until the seventeenth century BC that the Hyksos seem to have brought the horse to the Nile valley from western Asia. Here it rapidly became naturalised; then, from the eighteenth dynasty on (1580–1350 BC) under the Tuthmosis and Amenhoteps, and especially in the nineteenth dynasty (1350–1205 BC) under the Ramesses and Setis, it became a highly valued domestic animal.[20]

The horse was already drawing the war chariots in those days; it was also harnessed to the two-wheeled cars used for hunting. At the same time it was used in Babylonia and Assyria for war and hunting. Only centuries later does it seem to have been employed for riding. No African people – except for the Egyptians – possessed the horse then. Herodotus reports that the Arabs fighting in Xerxes's army rode on camels, 'that did not yield to the horses for speed.'

It is not until early Greek and Roman times that the horse is first used for riding. Now the human being swings up on to its back; now he becomes master of the horse. The magnificent relief on the Parthenon, which portrays boys riding in the Panathenian procession, marks the accomplishment of this mastery of the horse by man.

Even before the walls of Ilium, Greeks and Trojans fight not on horseback, but in horse-drawn chariots.*

All the facts gathered here testify to the horse's late assimilation into the cultural history of post-Atlantean humankind. It is not before the middle of the second pre-Christian millennium that

---

* The most beautiful and even today the most valuable description of the domestication of horses is found in Victor Hehn's famous work. We are told: 'Apart from war, in Homer the horse is not used for riding. This is evident from the third book of the *Odyssey,* where Telemachus and Nestor's son, Pisistratos, travel from Pylos to Lakedaemon across the mountainous Peloponnesos *standing in a chariot,* and not *riding* through the mountain passes or across the stony beds of mountain streams.'[21]

the horse is harnessed to the chariots of hunters and warriors; and not for still another millennium does man mount it and become a rider.

At the very same time the Mongols were storming across the Asiatic steppes on their little horses; and it is quite likely that the Germanic tribes settling more to the west adopted their use of the horse for hauling and riding. On their migrations to the north of Europe, they encountered the indigenous wild horse and tamed it. From their mythology it is evident that they perceived their gods as mounted beings: Wotan (Odin) and his hosts charge about on divine horses. In their cult they sacrificed primarily horses on the altar; in southern Sweden, a horse skull was found with the remains of a dagger still lodged in the frontal bone.

Perhaps we may assume that there were two streams on which the horse became part of human history. From the eastern, the wild horses that grazed the Asiatic steppes came across Turkestan, Persia and present-day Afghanistan to Babylonia and Assyria. These became the light-footed horses of war and riding. On the other, in northern Europe – in Scandinavia and northern Russia – the native wild horse was tamed, used as a sacrificial animal, and gradually became the beloved dray-horse *(Gaul)* and member of the family among the Teutons.

Reinhardt arrives at a similar assumption:

> The oriental or 'warm-blooded' horse, standing closer to the ass in skull structure, is the progenitor of all fleet-footed horses used for riding and drawing carriages. The less graceful but stronger occidental, 'cold-blooded' horse is the forefather of the heavy German carthorse, whose ancestors bore the medieval knights and their armour – heavy for both man and beast.[22]

Why did humans mount the horse so late? Various answers can be given. In prehistoric times, the horse was a sacred animal.

In India and Persia it was associated with the gods. Indra rode through the heavens on a horse; it was horses that drew the sun-chariot, and horses that pulled the moon through space. Should the earthly counterparts of these sacred beings be mounted by humans? They could only be offered as a holy sacrifice to those who employed them in the higher worlds.

However, as clairvoyance gradually dwindled in humankind, and in its place cognition awoke, horses were tamed, bridled, and harnessed to wagons. Their power was turned to human service. And later still, as the temples of the mysteries gradually closed their gates and the light of logic lit up in Socrates, Plato and Aristotle, the time came for the human being to sit astride the horse and become its lord and master.

Rudolf Steiner referred many times to such a path of human development. In the introduction to his *Riddles of Philosophy* we read:

> It is in Greece that the aspiration is born to gain knowledge of the world and its laws by means of an element that can, in the present age, be acknowledged as *thought*.[23]

And in *Universe, Earth and Man*, he says:

> We have seen that a handful of men (these were the proto-Semites who were led out of old Atlantis under the guidance of Manu) who dwelt in the neighbourhood of present-day Ireland ... had acquired the qualities that became manifest gradually during the succeeding cultural epochs ... These consisted in a disposition for logical thinking, for judgment. Before this, no such thing had existed; if a thought was present, it was already confirmed. The germ of discriminating judgment was planted in this handful of people, and they brought it over from the west to the east. Through their colonising

migrations, one of which went south to India, logical thought was introduced into the Persian culture. In the third culture-epoch, the Chaldean, this logical thinking grew even stronger, until the Greeks developed it into the glorious monument of Aristotelian philosophy.[24]

Thus we have here a view of how thought unfolded in the course of humankind's development; and in it we can recognise simply the inner aspect of what occurred outwardly in the gradual meeting of human being and horse.

The handful of proto-Semites which took form during the later Atlantean epochs, arose at the same time as eohippus, mesohippus, and orohippus. Both the proto-Semites and the hipparion migrated eastward, abandoning the drowning continent.

These people became the founders of the post-Atlantean cultures. Meanwhile, the horses stayed in their vicinity, waiting for closer contact. Invisibly, they carried within them the powers of thought which were to awaken slowly in humankind. It was the shadows of these cosmic thoughts on which the seers perceived the gods Indra, Wotan, and Poseidon riding.

Gradually, the human head became the bearer of thought. In Greece, this process came to perfection; the horse became a steed, and the human being a rider – a knight.

Girded with Aristotelian logic, he treads God's earth as master. Alexander, as Aristotle's pupil, can tame Bucephalus; for he put both horse and logic to his service. He became humankind's first knight.

Mythical horses

The Greek world of gods and heroes is rich in equine figures. In the most varied forms they appear alongside mortals and immortal beings. From the depths of the earth up to the heights of heaven, they fill the motley world of earthly deeds and supersensible events. And always it is a matter of overcoming the wildness, the great power of the horse through human cleverness and strength. The struggle between humans and ungulates – from Poseidon's wild arts of transformation to the cunning of Odysseus – stand at the centre of Greek horse mythology. This was perceived by Bachofen:

> The horse is an image of the untamed elemental power that reigns in the swamp, the power that wantonly scatters its fertilising seed over the earth ... and also a symbol of adulterous life. The reining in of this wild horse is the doing of woman.[25]

This wild, unruly being that tramples the earth with its hooves and stirs up the water of the seas with the fire of its passion, is subject to Poseidon. He himself took on the form of a stallion and approached the Earth-mother Demeter, who allowed him to mate with her. She bore him a daughter, whose name was not to be uttered outside the mysteries. Also from her womb sprang the famous black-maned horse Arion, who was as fast as the wind.

Poseidon was as close to the horses as Demeter was to the grain; and from the time he wedded Amphitrite and became ruler of the seas, there appeared these 'horse-monsters – half horse and half serpent-like fish – sea centaurs, whose animal lower bodies were a combination of horse and fish, Oceanides and Nereides, with names revealing their female equine nature: names such as Hippo, Hipponoe, Hippothoe and Manippe.'[26]

Manifested in Poseidon himself and in his hosts, we see that stormy equine nature which Bachofen speaks of as the 'water-power that wantonly inseminates the earth.' This means the meteorological forces, which seem to work without law and are also at work within a human beings whenever drives and passion pervade them.

Rudolf Steiner described the realm of Poseidon's power as the ancient Greek felt it:

He [the Greek] felt that in the ebb and flow of the
ocean, in the storms and hurricanes which rage over
the earth, the same forces are active as are active in us
when lasting emotions, when passion and habit pulsate
through our memory ... they are the forces in us which
we cover by the term 'ether body' ... The ancient Greek
was still conscious of a figure who could be reached
by clairvoyance, he was still conscious of the ruler, the
centre of all these forces in the macrocosm, and spoke of
him as Poseidon.[27]

The Nereides and Oceanides represent unclarified, unpuri-
fied, wild formative forces. They are seeking to be raised up and
transformed. The mythology of the Greeks points to this.

Poseidon's son, Bellorophon – a grandson of Sisyphos –
desired a winged horse. His father gave him the immortal
Pegasus. Pegasus, however, was also the son of Poseidon and
thus brother to Bellorophon. In primordial times, when the
Gorgons were still radiant in beauty, Poseidon loved one of
them, Medusa. For a long time she concealed the growing offspring in her womb; and not till Theseus beheaded her did the
winged Pegasus spring from her bleeding neck, Thus the horse
rose out of the darkest depths up to the light of day.

Bellorophon, a mortal hero, was at first unable to tame his
brother, Pegasus. However, Athena endowed him with a golden
charm as he begged her for help at one of her altars. 'The hero
mounted the divine steed and danced with it the war-dance in
full armour, to the goddess' honour.'[28]

So the horse, sprung from the depths, has been tamed and
mounted by a man with the help of the goddess of wisdom, who
gave him the 'golden' charm.

With the power conferred on him he performed deeds reminiscent of Herakles (Hercules). His victories – he also fought
against the Amazons – were so great that he won the daughter of
the King of Lycia to wife.

Then, however, the power of the newly awakened flight of thought takes revenge. Doubts arise in him: 'Are there really gods at all?' He wishes to find this out for himself and mounts Pegasus to ride up to Olympus, where he intends to penetrate into the council of the gods. This is too much. The heavenly horse throws off his misguided rider. He plunges to earth, and there – on the plain of Aleion (the plain of 'aimless wandering') – he leads a limping existence. Thus do doubts and misgiving in thought plunge us from the heights of recognition into the abyss of unknowing.

It was Herakles who became the real master of the horse. He tamed the horses of Diomedes; it was the seventh of the twelve deeds he had to perform. These beasts, like Pegasus, were winged, and were also related to the Harpies, Gorgons, and Erinyes: they ate human flesh. Their master, Diomedes, was a son of Ares. Herakles slew him and threw his body to the horses. In this way, they became tame and he was able to take them to Mycenae. It is told that Alexander's horse Bucephalus stemmed from their line.

Another aspect of the mythical horses is handed down in the centaur stories. Their origin is attributed to Hera herself, who, before she belonged to Zeus, was fertilised by giants – and perhaps by other primordial forces as well – and gave birth to centaurs. These creatures, half horse and half man, are in the process of transforming their animal nature into a human one. One of them – Chiron – became the teacher of Asclepios, as he still possessed divine wisdom. He knew magic as well as the healing power of herbs. Sometimes he is portrayed playing the lyre. Was he a pupil of Orpheus?

Rudolf Steiner spoke of the image of the centaur in connection with Gilgamesh. He describes this personality as indwelt and guided by an archangel, and goes on to say:

> Such a being did indeed work through Gilgamesh ... So we shall rightly conceive of this Gilgamesh in a picture

that suggests the symbol of the ancient centaur ...
A centaur – half-man, half-animal – was always intended
to represent how in the more mighty of men of old
the highest spiritual manhood and that which united
the single personality with the animal organisation in
a certain sense actually fell apart. Gilgamesh gave the
impression of a centaur to those who were capable of
judging what he was.[29]

This then is the manifold picture of the horse's mythical nature we have received from the Greek world. It permeates through and through the fading imagery of that age, and sets human beings the task of meeting their own nature. Athena. Herakles, Bellorophon, the centaurs, and Poseidon: all of them appeared in plays and dramas which provided an exoteric guidance towards the mysteries.

In these, it was taught how human nature had expelled horse-nature from itself in past times, so that the human being could be endowed with the power of intelligence. This secret was revealed as the fifth page of the 'ten-page book'. The opening of this fifth page is presented in the Revelation to St John. Here, the Lamb unseals the first four seals of the Book of Life. 'And I saw, and behold a white horse' (Rev.6:2). This is followed by a red horse; then comes a black horse, and lastly a pale horse. Each horse has its rider, who bears particular gifts. The rider of the white horse wore a crown, 'and he went forth conquering, and to conquer.' The rider of the red horse is given a great sword. 'to take peace from the earth.' The master of the black horse 'had a pair of balances in his hand.' On the pale horse rides he whose 'name ... was Death, and Hell followed with him.'

In a detailed discussion of these images, Rudolf Steiner clarifies their meaning. The victory procession described here is that of the growing power of the intellect. The four post-Atlantean ages appear in the form of the four horses and those who ride upon them.*

> By having passed through the Indian age in a frame
> of mind in which he turned away from the physical
> world and directed his gaze towards the spiritual, man
> will, in the first age after the War of All Against All,
> gain the victory over the things of sense. He will be
> the victor by acquiring what was written in his soul in
> the first age. Further what emerged in the second age,
> the conquest of matter by the ancient Persians, will
> appear in the second age after the War of All Against
> All: the sword that here signifies the instrument for the
> overcoming of the external world. What man acquired
> in the Babylonian-Egyptian age, when he learned how
> to measure everything correctly, is seen in the third age
> after the great War as that which is represented by the
> scales. And the fourth age shows us something most
> important, namely what man acquired in the fourth age
> of our epoch through Christ Jesus and his appearance on
> earth: the spiritual life, the immortality of the ego. For
> this fourth age it must be shown that everything unfit for
> immortality, everything doomed to die, falls away.[31]

Here the true mission of the horse in human history becomes clear again. Through its sacrifice we human beings have acquired our intellect. What remained as natural power and might in the horse is transformed in the human being into the life of thought.

---

* The post-Atlantean epoch can be divided into seven ages. It began in 7900 BC and ends 15,120 years later with the War of All Against All. In his lectures on the Apocalypse, Steiner says: 'Then [the I] becomes what separates men from one another, what brings them to the great War of All Against All, not only to the war of nation against nation (for the conception of a nation will then no longer have the significance it possesses today) but to the War of each single person against every other person in every branch of life; to the war of class against class, of caste against caste, and sex against sex.'[30]

In this connection, introducing the above exposition, Rudolf
Steiner said:

> Curious and grotesque as it may appear, it is nevertheless
> true to say that if there were not around us the animals
> which belong to the horse nature, man would never have
> been able to acquire intelligence.
>
> In former times, people were still aware of this. All
> the intimate relation existing between certain human
> races and the horse originate from a feeling which may
> be compared to the mysterious feeling of love between
> the two sexes, from a feeling of what man owes to this
> animal. Hence when the new culture arose in the ancient
> Indian age, it was a horse that played a mysterious role
> in religious ceremonial, in the worship of the gods.
> And all custom connected with the horse may be traced
> back to this fact. If you observe the customs of ancient
> peoples who were still close to the old clairvoyance, the
> old Germanic peoples, and notice how they fixed horse-
> skulls to the front of their houses, this leads you back to
> the awareness: man has grown beyond the unintelligent
> condition by separating out this form. There is a
> profound consciousness that the acquisition of cleverness
> is connected with it.[32]

Epilogue by Fritz Götte

The word 'epilogue' is the last one still written by Karl König in
his work 'Brother Horse'. It stands in his striking handwriting, as
though lost, on page 48 of his manuscript. At midday on March
27, 1966, death took the pen from this tireless writer's hand.

In the discussion we had about his work, König remarked that
the Apocalypse of John would play a substantial part in it. I hope
that what I have added is not completely astray. König saw in the

redemption of human intelligence the decisive service which the human being may give to his 'brother horse'. That redemption is presented in the Apocalypse as the image of the Word of God riding on the white horse.

All that König left us are some notes which at most allow us to divine in what way he wanted to crown his work. There stands the following:

> *The musicality of the horse nature!*
> *It lives on further – without having the possibility of being transformed.*
> *Orpheus, son of Apollo! Michael!!*
> *It was music which had an effect on the physical plane; but the sentient soul had the unconscious feeling:*
> *That comes out of regions from where derives the light. Music, song, out of the realm of light (first lecture from Rudolf Steiner's 'The Christ-Impulse').*[33]
> *But this realm is the world of the sun.*

We know further how keen he was that the picture by Hans von Marées, brought here as an illustration, should be added to his essay. For it belongs, so he said, among this work's initial experiences.

Then there is still a passage from the book by Julius Schwab which he noted down:

> In Vedic mythology the sun is called Stallion, in another place a 'steed with seven names'. The dawn is said to be led by a white steed: the sun. The sacrificial steed is said to be produced by gods out of the sun. In Hellas people sacrificed on a peak of the Taygetos, sacred to Helios, among other things also horses. At the temple in Jerusalem there were set up steeds consecrated to the sun, removed by Joshua (2Kings 23). Sigurd's steed Grani, which according to an old Danish song can climb

on its own the glass mountain (mountain of the sky), is
the sun ...

There can be no doubt: towards the end of his work König
was reaching out towards the secret of the sun, that works in
human intelligence in hidden ways, that is to say in its connection with the horse. It is his actual theme, from which he justified adding the word 'brother' to the word 'horse'. And indeed:
can there be a deeper relation between animal and humans than
one relating to what raises the human being above creation: the
power of thought?

Earlier periods of history called the beings of the higher hier-
archies from Angels up to Thrones, Seraphim and Cherubim
'intelligences'. There have always been spiritual streams which
took these intelligences of the world to be apprehensible precisely through the power of reason. And out of anthroposophy
it can be learnt that this cosmic, not yet earthly, intelligence was
the responsibility of the Archangel Michael.

Hans von Marées, Rider Triptych,
St Martin, St Hubertus, St George

Victor Hehn, to whom Karl König also refers, gives numerous examples of the horse being experienced and treated in ancient cultures as an animal related to the sun. As evidence we mention only Herodotus, who said of the Massagetae: 'As God they worship only the sun, to whom they sacrifice horses.' And Herodotus adds in explanation: 'The meaning of this sacrifice is as follows: to the fastest of all gods they allocate the fastest of all earthly creatures.' Stormy, storm-footed and similar descriptions are found in Greek names for the horse, and Hehn says, supported by his many sources: 'The steed melts in vision into the storm.' Achilles's steeds were 'sons of Zephyr', and whoever has read all these passages would actually like to say: horse and wind are fast; they are fast as thought, which for its part is like sunlight. And it is also steeds which in Aeschylus and Sophocles bring the sun's day onto earth.

'With the Germanic peoples,' so Hehn elaborates, 'the cult devoted to the steed also bears some quite Iranian traits.' But the Iranian culture of Mazda was indeed through and through a service of the sun. Yet another side of the higher equine nature can also be considered when Hehn continues: 'Horses possess the power of prophecy, they are sacrificed to the gods, they draw the sacred chariot' (which, as may be added, bears the picture of a god, a hierarchical intelligence). And Otto Schrader, the great linguist, bluntly describes the horse as an 'oracle animal', as such playing an important role for the Persians (Herodotus), for the Germans (Tacitus) and for the West Slavs.[34] Thus from this side as well, the horse's intelligent nature is made evident from mythology. But Otto Schrader also refers to the Fall, which the human being allowed to be committed towards the horse. He says, above all by reason of his linguistic studies:

> A great change in the European terminology of the horse is brought about by the animal's social position, which changed from the animal being used since prehistoric times at first only for sacred purposes, and then especially in war, the *bellator equus,* to being pressed more and more into the general service of human beings.

It was precisely this phenomenon that was important for Rudolf Steiner when speaking in his lecture-cycle on the Apocalypse of St John about the appearance of the four horses at the opening of the seals. He sees the development of human intelligence bound up with the nature of the horse, and the opening of the apocalyptic seals, relating to the four post-Atlantean cultures, indicate stages on this downward path of the intelligence from cosmic heights into earthly development. The fifth post-Atlantean culture – that is our own – must bring a change: the raising of the earthly (and thereby egoistically mis-used) intelligence to a selfless one, which becomes appropriate to the nature of the world, and no longer harms and violates creation. The Apocalypse describes this change when the open-ing of the fifth seal is no longer accompanied by the horses as in the first four cultures, but by human beings in white garments.

Rudolf Steiner called the battle for the fate of cosmic intelligence in our time a Michaelic struggle. On the one hand there is the danger of sinking into ever greater depths, and on the other hand there is the possibility of rising up to the Logos, out of which everything was created that was created.

In his lectures on the Philosophy of Thomas Aquinas, Rudolf Steiner formulated the great decisive question as follows:

> How can thinking be made Christian? ... This question stands there in world history at the moment in 1274 when Thomas Aquinas dies. Up to that moment he could only struggle through to the question ... The question stands there with all inwardness of heart in European spiritual culture.[35]

The human being has in course of history become a rider, as König implied. The human being makes use of the horse, or we can also say, he uses the originally cosmic intelligence and gradually places it more in the service of his earthly and egoistic needs, forgetting his own heavenly origin. But the material and technical manner of thinking into which he has transformed his own (now personal) intelligence also had the consequence that the being of the horse in its animal form noticeably vanished once again from our culture, merely leaving the faint shadow of h.p., the horsepower of present-day engines. It is an even deeper stage in the development Otto Schrader described. Today what Rudolf Steiner said in his lectures on the Apocalypse holds true in regard to western technical and intelligent culture:

> This intelligence will spread. People will obtain what is needed for their body with even much greater powers of mind. They will kill each other with much greater powers of mind until the great War of All against All sets in. Many discoveries will be made in order to wage wars

more effectively. Endless intelligence will be mustered in order to satisfy the lower desires.

With the desecration of the horse, we could say that the human being has debased himself by desecrating his intelligence. But in the process Christianity must sink as well. Its fate will be determined by the answer found to that question to which Scholasticism could not reply: how will thinking itself be Christianised, how will thought-substance as such be made Christian? The quality of thought will determine what tools the future will bring: tools that annihilate fellow humans and nature, or ones bearing the essence of healing, able to pass before Christ. The quality of thought will also determine the social and political order or disorder humankind will find. These questions, as everyone can clearly see, are of an Apocalyptic dimension and can already be sensed.

To Rudolf Steiner the Pauline question about the salvation of creation encompassed, indeed was even headed by, human thinking. It was for Steiner a matter of overcoming what he called the 'dogma of experience' dominating science, which has taken the place of the church's 'dogma of the revelation'. Experience is limited somewhat arbitrarily and dogmatically to what can be grasped by the senses, measured and mathematically calculated.

The powerful influence that this manner of perception has on humankind today is a self-limitation that fetters the human soul and spirit. Humankind is challenged to break this limitation and tread new paths to acquire higher and greater cognition. This path will enable research and experience of the spirituality where the Logos holds sway, out of which our whole created world has emerged and been ordered.

Unredeemed human reason alone cannot rise into the spiritual world. Redeemed human reason, which has a real relation to Christ, penetrates into the spiritual

world. Penetrating into the spiritual world from this
point of view is Christianity of the twentieth century,
is a Christianity so strong that it penetrates into the
innermost fibres of human thinking and human soul-
life.[36]

That is what Rudolf Steiner said in the lectures on Thomas
Aquinas. He described the way towards it in his book *Knowledge
of the Higher Worlds*. He leads through the gradual overcoming
of egoism, to which human cognition has sunk, into the free
heights of the spirit, where the original cosmic intelligence holds
sway. Among them is the Archangel Michael, the selfless ruler
of the cosmic intelligence in the service of Christ, shining before
him as his countenance. By resanctifying our intelligence and
through this our entire selves, the human being becomes again
a legitimate rider, riding on the horse of purified and redeemed
intelligence. He strives towards the image of the rider and the
horse of the Apocalypse.

Then I saw heaven opened, and behold, a white horse!
He who sat upon it is called Faithful and True, and in
righteousness he judges and makes war. His eyes are like
a flame of fire, and on his head are many diadems; and he
has a name inscribed which no one knows but himself.
He is clad in a robe dipped in blood, and the name by
which he is called is The Word of God. (Rev.19:11–13).

Publication details

The essays in this volume were first published in the journal *Die Drei* in Stuttgart, starting with 'The Migrations of Salmon and Eels' and 'The Dove as a Sacred Bird' in 1956.

'The Origin of Seals', translated by Richard Aylward, first published in English in *Penguins,* Floris Books 1984.

'The Life of Penguins', translated by Richard Aylward, first published in English in *Penguins,* Floris Books 1984.

'The Migrations of Salmon and Eels', first published in English in *British Homoeopathic Journal,* July 1962.

'Dolphins – Children of the Sea', first published in English in *The Golden Blade,* 1967.

'Swans and Storks', translated by Richard Aylward, first published in English in *Swans and Storks,* Floris Books 1987.

'The Dove as a Sacred Bird', first published in English in *The Golden Blade,* 1968.

'The Sparrows of the Earth', translated by Richard Aylward, first published in English in *Swans and Storks,* Floris Books 1987.

'Elephants', translated by Richard Aylward, first published in English in *Elephants,* Floris Books 1992.

'The Bear Tribe and its Myth', translated by Richard Aylward, first published in English in *Elephants,* Floris Books 1992.

'Cats and Dogs – Companions of Man', translated by Richard Aylward, first published in English in *Elephants,* Floris Books 1992.

'Brother Horse', translated by Richard Aylward, first published in English in *Elephants,* Floris Books 1992.

Notes

Professor Wolfgang Schad and Dr Johannes F. Brakl have kindly provided some of these notes with updated zoological research for this edition in 2013.

Introduction

1 Bertha König, *Meine Kindheits- und Lebenserinnerungen.*
2 After 18.6 years (about 18 years, 7 months and 9 days) the position of the moon's path in relation to the sun's path around the earth comes to the same point against the background of the stars. In many biographies this period marks a special time; for König this period influenced the rest of his life.
3 König's diary, quoted in Selg, *Karl König: My Task,* pp. 62f.
4 Diary quoted in Müller-Weidemann, *Karl König,* p. 32.
5 Selg, *Karl König: My Task,* pp. 18f.
6 Charles Darwin, *Autobiography,* p. 112f.
7 Selg, *Karl König: My Task,* p. 21.
8 Steiner, *Founding a Science of the Spirit,* lecture of August 30, 1906.
9 After a drawing by Rudolf Steiner in *Fundamentals of Esotericism,* lecture of September 28, 1905.
10 Steiner, *An Esoteric Cosmology,* lecture of May 26, 1906, p. 9.
11 Goethe, *Tag und Jahreshefte,* 1790.
12 König, *Die Ordnung der Tiere im Tierkreis,* 3 booklets.
13 König, *The Animals and their Destiny.*
14 About this faculty of imaginative thinking see Richard Steel, 'This Land is One That We Can Now Intuitively Sense,' in König, *Kaspar Hauser and Karl König.*
15 Steiner, *Macrocosm and Microcosm,* lecture of March 31, 1910.

1. The Origin of Seals

1 Lucanus, *Zugvögel und Vogelzug.*
2 This only applies to the common harbour seal. Eared seals and sea lions are weaned for about a year, until the birth of the next young. Also only the common seal changes the milk teeth so early (H.

Schliemann in Grzimek, *Enzyklopädie,* Vol. 4).

3 This is true of all eared seals and sea lions, the harbour seal does not seem to have such a clear social structure.

4 The oldest fossil found to date is an ancestor of the monk seal from the Middle Miocene. Predecessors of the common seal have not been found. (E. Thenius in Grzimek, *Enzyklopädie).*

5 Scheffer, *Seals, Sea Lions and Walruses.*

6 Some of the seals living in inland waters are today considered to have moved there from the open sea during the ice-age, as König already considered (E. Thenius in Grzimek, *Enzyklopädie,* Vol. 4).

7 Steiner, *Egyptian Myths and Mysteries,* lecture of Sep 7, 1908, pp. 54f, 56.

8 Wachsmuth, *Entwicklung der Erde.*

9 Steiner, *Cosmic Memory,* Ch. 3.

10 König considers the development of the human form in a spiritual process that took place in the supersensible realm, not in the physical world, whereas animals such as the seals had already become part of material evolution and therefore no longer had a part in human development.

11 Frazer, *The Golden Bough,* Vol. 3, p. 210.

12 Steiner, *Geisteswissenschaftliche Erläuterung,* Vol. 2, lecture of Nov 2, 1917.

13 For the text of this story, Karl König was indebted to Professor Kohl-Larsen who heard it with him and took it down.

2. The Life of Penguins

1 Kearton, *Island of Penguins,* pp. x-xi.

2 Today it is assumed that they seek to keep in a constant temperature zone because the layers of fat as well as their highly insulating down-feathers do not withstand fluctuating temperatures.

3 Banse, *Geographische Landschaftskunde.*

4 Hermann, *Die Pole der Erde.*

5 Brehm, *Vögel.*

6 Gerlach, *Die Gefiederten.*

7 The last two surviving specimens were caught and killed in 1844 off the south-west coast of Iceland.

8 Brehm, *Vögel.*

9 Portmann, *Von Vögeln und Insekten.*

10 See Marret, *Sieben Mann bei den Pinguinen,* and Rivolier, *Emperor Penguins.*

11 Kearton, *The Island of Penguins,* p. 83.

12 The structure of the wings indicate this as all the bones of a wing suitable for flight are present. These are, however, shorter, flatter and connected by ligaments forming a firm surface like flippers. (B.

Stonehouse in Grzimek, B. *Enzyklopädie,* Vol.7.)

13 Kearton, *The Island of Penguins,* pp. 89, 93, 97f.

14 Steiner, *Menschenwerden, Weltenseele und Weltengeist,* Vol. 1, p. 91 (lecture of July 1, 1921).

3. The Migration of Salmon and Eels

1 *Coelecanthidae* were filmed for the first time in 1987, at night from the German research submarine *GEO.* Three years later a group of 10 or 15 was found at a depth of 400 metres (1,300 feet) off the steep West coast of Anjouan, Comoros Islands. They hid during the day in lava caves but went out on the hunt at night in depths of around 700 metres (2,300 feet). In 1998 a 1.2 metre (4 ft) long coelacanth weighing 30 kg (66 lb) was caught in a net near Sulawesi, Indonesia over 8,000 km (5,000 miles) from the only habitat previously known, at the Comoros Islands. Unlike the steel-blue white-spotted Comoros species, it was a chocolate colour and probably belonged to a different species. (Kleesattel, *Die Welt der lebenden Fossilien.)*

2 Quoted in Brehm, *Die Fische.*

3 Brehm, *Die Fische.*

4 Roule, *Fishes,* pp. 204f.

5 Gerlach, *Die Fische.*

6 Steiner, *From Elephants to Einstein,* lecture of Feb 9, 1924.

7 Steiner, *From Elephants to Einstein,* lecture of Feb 9, 1924.

8 It is now known through many experiments that salmon find their home river by their sense of smell. In fact they do not remember the smell or taste of their river at the time of their birth but at the time when they *left* it. In check-studies the nose indentures of salmon (which they do not need for breathing) were blocked up which caused the fish looking for their origin to distribute equally amongst all of the higher situated streams. The other salmon all found their way back to the exact stream of their birth, which is an amazing feat considering the measure of dilution of that smell being remembered and found! Karl König's indication of a certain light perception means etheric light and the light-ether that is present in any sensory perception.

9 Roule, *Fishes,* p. 200.

10 Steiner, *Harmony of the Creative Word,* lecture of Oct 28, 1923, pp. 110f.

11 Muir-Evans, *Sting-Fish and Seafarer,* p. 87.

12 Steiner, *The Mission of Folk-Souls,* lecture of June 16, 1910, pp. 161f.

13 Steiner, *Universe, Earth and Man,* lecture of Aug 11, 1908, pp. 100f.

14 *Latimeria,* the coelecanth, is significantly older than most other boned fish: Fossils can be found as far back as the Devonian period. It may seem to be a contradiction that the coelecanth is the ancestor of amphibians and therefore of all land vertebrates. The direct ancestors

were however not today's factually almost unchanged *Latimeria* but probably the closely related *Rhipidistia* (Kleesattel, *Die Welt der lebenden Fossilien*).

15 Steiner, *Universe, Earth and Man,* lecture of Aug 11, 1908, p. 104.

16 It is important to note that König means the *form* of the snake and not the snake itself as an evolutionary predecessor of the reptiles.

17 Brehm, *Die Fische.*

18 This can be seen by the difference in behaviour according to whether the moon is full or new (Endres & Schad, *Moon Rhythms in Nature*).

19 Steiner, *Universe, Earth and Man,* lecture of Aug 11, 1908, p. 104.

4. Elephants

1 Steiner, *Antworten der Geisteswissenschaft,* lecture of March 10, 1910, p. 69.

2 There are some exceptions to this generally true statement. The Carthaginians used the now extinct North African elephant as a working animal. The Belgians set up an elephant training station in their former colony of the Congo; this was in use until 1963 and may be restarted. The African elephant is tameable although much more difficult, and therefore it is seldom done. (Grzimek, *Enzyklopädie,* Vol. 12, and Douglas-Hamilton, *Battle for the Elephants*).

3 Steiner, Steiner, *Antworten der Geisteswissenschaft,* lecture of Nov 10, 1910.

4 It is now known that the visual acuity is similar to that of a horse but it may well be that its significance recedes due to the fact that the direction of sight with normal head posture is to the ground and sideways. To look forwards the elephant has to first lift it head.

5 Studies done by Katy Payne show that an elephant's hearing encompasses an infrasonic region inaudible for the human ear. These are extremely low sounds with which elephants can communicate to each other over a distance of some miles. Infrasonic sound is also produced by wind and waves and is transported by the ground. In this way elephants are gifted by an extremely comprehensive perception of earth, water and air.

6 Carrington, *Elephants,* p. 61.

7 Blond, *The Elephants.*

8 Steiner, Steiner, *Antworten der Geisteswissenschaft,* lecture of Nov 10, 1910.

9 Carrington, *Elephants,* p. 62.

10 According to Katy Payne's studies elephants can also recognise individual voices from a great distance.

11 Even though *Moeritherium* is no longer considered to be predecessor of the present day elephant but of an extinct branch line, it is nevertheless the earliest form of the proboscidea and connecting link to the sea cow (E. Thenius in Grzimek, *Enzyklopädie,* Vol. 4).

12 Carrington, *Elephants,* p. 108.
13 Carrington, *Elephants,* p. 70.
14 Blond, *The Elephants.*
15 See Steiner, *Cosmic Memory.*
16 Steiner, *East in the Light of the West.*

5. The Bear Tribe and its Myth

1 Brehm, *Die Säugetiere.*
2 Although it has extended its territory, the numbers of the Asiatic black bear has receded today quite strongly and also the sloth bear has almost been made extinct by the clearing of its habitat, the Indian dry woodlands; at present it is only to be found in the National Parks.
3 Abel, *Paläobiologie.*
4 Poppelbaum, *Tierwesenskunde.*
5 A. Pedersen writes: 'According to a legend of the Shasta tribe, Manitou created the grizzly as a more mighty and clever being than any others. At that time the grizzly walked upright like the human being did later on. It did not kill its prey with its teeth and claws but with a club.' (In Grzimek, *Enzyklopädie,* Vol. 12,).
6 Brehm, *Die Säugetiere.*
7 Gerlach, *Die Vierfüssler.*
8 Frazer, *The Golden Bough,* p. 506.
9 Frazer, *The Golden Bough,* p. 315.
10 Bayley, *The Lost Language of Symbolism,* Vol. 1, p. 115.
11 Frazer, *The Golden Bough,* p. 221.
12 Kerényi, *The Gods of the Greeks.*
13 Schadewaldt, *Die Sternsagen der Griechen.*
14 Fuhrmann, *Das Tier in der Religion.*
15 See Hoenn, *Artemis.*
16 Steiner, *Cosmic Memory,* pp. 53, 55.
17 Steiner, *The Karma of Untruthfulness,* Vol. 1, lecture of Dec 21, 1916.
18 Bernatzik, *Die neue grosse Völkerkunde.* Vol. 2.
19 Gleich, *Der Mensch der Eiszeit und Atlantis.*
20 Steiner, *Wonders of the World,* lecture of Aug 23, 1911, p. 113.

6. Swans and Storks

1 Steiner, *Harmony of the Creative Word,* lecture of Oct 27, 1923. pp. 89f.
2 Gerlach, *Die Gefiederten.*
3 See Makatsch, *Die Vögel der Erde.*
4 Steiner, *Wonders of the World,* lecture of Aug 26, 1911, p. 157.
5 Steiner, *The Gospel of St John and its Relation to the Other Gospels,* lecture of July 1, 1909, p. 145.

6 Only the whistling swan migrates as a family group to its winter quarters.

7 Siewert, *Störche.*

8 The Abdim's stork for instance wanders within Africa. It spends the winter in the savannas of Southern Africa and moves back to the savannas of the northern half of Africa only at the start of the rain season and is celebrated as the rain-bringer as well as the baby-bringer, whereas our white stork is not respected there and even gets hunted. (Curry-Lindahl, *Vogelzug,* and Schulz, *Weissstorchzug.*)

9 Steiner, *Die okkulten Wahrheiten,* lecture of Dec 3, 1905.

10 Steiner, *The Mystery of the Trinity,* lecture of July 23, 1922, pp. 13f.

7. The Dove as a Sacred Bird

1 Brehm , *Vögel.* Vol. 1.

2 *The Gospel of St Matthew,* lecture of Sep 1, 1910, pp. 30f.

3 A similar function has only been found with flamingos. Leslie Brown discovered their breeding grounds only in 1958 but the breeding process was not studied until much later, when it was found that they feed their young on a similar fluid, although it is mixed with blood and therefore a red milk (C. Willcock, 'The African Rift-Valley' in *Time-Life,* 1978/13).

4 Steiner, *Harmony of the Creative Word,* lecture of Oct 27, 1923, p. 85.

5 Steiner, *The Gospel of St John and its Relation to the Other Gospels,* lecture of July 3, 1909, pp. 184f.

8. The Sparrows of the Earth

1 These descriptions and others in this study are based on Summers-Smith, *The House Sparrow.*

2 Bachelard, *Poetik des Raumes.*

3 Makatsch, *Die Vögel der Erde.*

4 Brehm, *Vögel,* Vol. 4.

5 Summers-Smith, *The House Sparrow,* p. 209.

6 Quoted in Summers-Smith, *The House Sparrow,* p. 176.

7 Summers-Smith, *The House Sparrow,* p. 209.

9. Dolphins – Children of the Sea

1 Alpers, *A Book of Dolphins,* p. 136f.

2 *Tier,* Hallwag, Berne Jan 1964, Vol. 3, No. 1.

3 Rabinovitch, *Der Delphin in Sage und Mythos der Griechen,* and Hildegarde Urner, 'Der Delphin als religions- und kunstgeschichtli-

ches Motiv'. *Neue Zürcher Zeitung*, Oct 3, 1959.

4 Alpers, *A Book of Dolphins,* and Slijper, *Whales.*

5 Slijper, *Whales,* p. 274.

6 Lilly, *Man and Dolphin,* p. 36.

7 Norris, Kenneth S., John H. Prescott, Paul V. Asa-Dorian, and Paul Perkins, 'An Experimental Demonstration of Echo-Location Behaviour in the Porpoise, *Tursiops truncatus,.' Biological Bulletin,* Marine Biology Laboratory, Woods Hole, Mass. April 1961.

8 Lilly, *Man and Dolphin,* p. 73.

9 In the meantime one knows that they can at least distinguish between the main taste *qualities*; sour, salty, bitter and sweet (W.Gewalt in Grzimek, *Enzyklopädie,* Vol. 4).

10 Slijper, *Whales.*

11 Lilly, *Man and Dolphin,* p. 36.

12 This assumption of König's has been proved correct. Bercovich writes that dolphins only sleep with half of their brain; the other half stays awake so that they can still swim and surface for air.

13 Steiner, *The Boundaries of Natural Science,* lecture of Oct 3, 1920.

14 Slijper, *Whales,* p. 204.

15 Kerényi, *The Gods of the Greeks.*

16 Steiner, *Wonders of the World,* lecture of Aug 22, 1911, pp. 98f.

17 Steiner, *Wonders of the World,* lecture of Aug 24, 1911, p. 120.

18 Steiner, *Christ and the Spiritual World,* lecture of Dec 30, 1913, p. 66.

19 Steiner, *Christ and the Spiritual World,* lecture of Dec 30, 1913, pp. 64, 68.

10. Cats and Dogs – Human Companions

1 See Wilckens, *Grundzüge der Naturgeschichte der Haustiere.*

2 See Reinhardt, *Kulturgeschichte der Nutztiere,* p. 6.

3 Schwabe, *Archetyp und Tierkreis,* p. 594.

4 In Brehm, *Die Säugetiere,* Vol. 3, p. 150.

5 In Brehm, *Die Säugetiere,* Vol. 3, p. 51.

6 Guggisberg, *Simba.* p. 20.

7 Portmann, *Die Tiergestalt,* Chapter 7.

8 *Goethes naturwissenschaftliche Schriften,* Vol.1.

9 Crisler, *Wir heulten mit den Wölfen.*

10 Suchantke, A. 'Was spricht sich in den Prachtkleidern der Vögel aus ?' *Die Drei,* 1964 No. 4.

11 Steiner, *Die Erkenntnis der Seele und des Geistes,* lecture of March 18, 1908.

12 Ash, *Dogs, their History and Development.*

13 Schwabe, *Archetyp und Tierkreis,* p. 95.

14 Erman, *Die Religion der Ägypter,* p. 33.

15 Erman, *Die Religion der Ägypter*, p. 42.
16 Steiner, *Wonders of the World*, lecture of Aug 26, 1911.
17 Wachsmuth, *The Evolution of Mankind.*
18 Steiner, *Cosmic Memory.*
19 Steiner, *Egyptian Myths and Mysteries*, pp. 79ff.
20 Steiner, *The Gospel of St Matthew*, lecture of Sep 1, 1910, pp. 33f.

11. Brother Horse

1 Oken, *Lehrbuch der Naturphilosophie.*
2 See Poppelbaum, *Tierwesenskunde.*
3 Hertwig, *Lehrbuch der Zoologie*, p. 616.
4 Today elephants, sea cows, aardvarks and the hyrax are also classed as hoofed animals, but within this strangely diverse group the odd-toed and even-toed animals represent two distinct sections.
5 See Georg Ritter, 'Horn und Geweih,' in *Kalender Ostern 1935 – Ostern 1936,* Mathematisch-Astronomische Sektion, Dornach 1935.
6 Poppelbaum, *Tierwesenskunde*, p. 104.
7 Brehm, *Die Säugetiere*, Vol. 3, p. 599.
8 Abel, *Paläobiologie und Stammesgeschichte*, p. 288.
9 Young, *The Life of Vertebrates*, p. 695.
10 Abel, *Paläobiologie und Stammesgeschichte*, p. 291.
11 This is in agreement with information given in Young, *The Life of Vertebrates.*
12 Steiner, *Universe, Earth and Man*, lecture of Aug 11, 1908.
13 Steiner, *From Comets to Cocaine*, lecture of Nov 29, 1922, pp. 43f.
14 The following description is based on the work of René du Bois-Reymond, 'Ortsbewegung der Säugetiere, Vögel, Reptilien und Amphibien,' in Bethe & Bergmann, *Handbuch der normalen und pathologischen Physiologie.*
15 Kluge, *Etymologisches Wörterbuch.*
16 Finbert, *Pferde.*
17 Reinhardt, *Kulturgeschichte der Nutztiere.*
18 Today (2013) the wild horses of Central Asia are considered to be extinct due to extensive hunting. Only the Przewalski's horse has survived in captivity. In Mongolia there are attempts at present to return zoo animals to the wilds.
19 Pritzwald, *Die Rassengeschichte der Wirtschaftstiere.*
20 Reinhardt, *Kulturgeschichte der Nutztiere.* p. 189.
21 Hehn, *Kulturpflazen und Haustiere.*
22 Reinhardt, *Kulturgeschichte der Nutztiere.* p. 203.
23 Steiner, *Riddles of Philosophy*, p. 6.
24 Steiner, *Universe, Earth and Man*, lecture of Aug 16, 1908.
25 Bachofen, *Das Mutterrecht*, Vol. 1, p. 125

26 Kerényi, *The Gods of the Greeks.*

27 Steiner, *Wonders of the World,* lecture of Aug 25, 1911, p. 53.

28 Kerényi, *The Heroes of the Greeks,* p. 81.

29 Steiner, *Occult History,* Dec 28, 1910, p. 44.

30 Steiner, *The Apocalypse of St John,* lecture of June 25, 1908, p. 137.

31 Steiner, *The Apocalypse of St John,* lecture of June 21, 1908, p. 86.

32 Steiner, *The Apocalypse of St John,* lecture of June 21, 1908, p. 82.

33 Steiner, *The Christ Impulse,* lecture of Oct 25, 1909.

34 Schrader, 'Pferd' in Vol. 2, *Reallexikon der indogermanischen Altertumskunde,* Berlin 1929.

35 Steiner, *The Redemption of Thinking,* lecture of May 23, 1920.

36 Steiner, *The Redemption of Thinking,* lecture of May 24, 1920.

Bibliography

Abel, Othenio, *Paläobiologie und Stammesgeschichte,* Fischer, Jena 1929.

Alpers, Antony, *A Book of Dolphins,* Murray, London 1960.

Ash, Edward C. *Dogs, their History and Development,* London 1927.

Bachelard, Gaston, *Poetik des Raumes,* Munich 1960.

Bachofen, Johannes Jacob, *Das Mutterrecht,* Basel 1861/1958.

Banse, Ewald, *Geographische Landschaftskunde,* Perthes, Gotha 1932.

Bayley, H. *The Lost Language of Symbolism,* London 1951.

Benedict, Francis G. *The Physiology of the Elephant,* Carnegie Institute, Washington 1936.

Bercovich, P,N, *The Dolphin's Boy,* London 2001.

Bernatzik, H.A. *Die neue grosse Völkerkunde.* Frankfurt am Main 1954.

Bethe, A. & G. von Bergmann, et al., *Handbuch der normalen und pathologischen Physiologie,* Berlin 1930.

Blond, Georges, *The Elephants,* Deutsch, London 1962.

Brehm, Alfred, *Tierleben,* Bibliographisches Institut, Leipzig 1911 (published in English as *Brehm's Life of Animals).*

—, *Die Fische,* Vol. 3 of *Tierleben.*

—, *Die Säugetiere,* in *Tierleben.*

—, *Vögel,* (4 volumes) Vol. 6—9 of *Tierleben.*

Carrington, Richard, *Elephants,* London 1958.

Carson, Rachel, *Silent Spring,* Houghton Mifflin, New York 1962.

Crisler, Lois, *Wir heulten mit den Wölfen,* Wiesbaden 1961.

Curry-Lindahl, K. *Vogelzug,* Berlin and Hamburg 1982.

Darwin, Charles, *Autobiography,* London 2005.

Douglas-Hamilton, Ian & Oria, *Battle for the Elephants,* Doubleday, London 1992.

Endres, Klaus-Peter & Wolfgang Schad, *Moon Rhythms in Nature,* Floris Books, Edinburgh 2002.

Erman, Adolf, *Die Religion der Ägypter,* Berlin 1934.

Evans, *see* Muir-Evans, Harold.

Finbert, Elian-J, *Pferde,* Zurich-Stuttgart 1963.

Frazer, James George, *The Golden Bough,* Macmillan, London 1911.

Fuhrmann, E. *Das Tier in der Religion,* Munich 1922.

Gerlach, Richard, *Die Fische,* Claassen, Hamburg 1950.

—, *Die Gefiederten: das schöne Leben der Vögel,* Droemersche, Munich & Zurich 1964.

—, *Die Vierfüssler,* Hamburg 1951.

Gleich, S. von, *Der Mensch der Eiszeit und Atlantis,* Stuttgart 1936.

Goethe, Wolfgang von, *Goethes naturwissenschaftliche Schriften,* (ed. Rudolf Steiner).

Grzimek, B. *Enzyklopädie der Säugetiere,* Droemersche, Munich 2000.

Guggisberg, C.A.W. *Simba, Eine Löwenmonographie,* Berne 1960.

Hamilton, *see* Douglas-Hamilton, Ian & Oria.

Hehn, Victor, *Kulturpflazen und Haustiere in ihrem Übergang aus Asien nach Griechenland und Italien sowie in das übrige Europa,* 1870.

Hermann, Ernst, *Die Pole der Erde,* Safari, Berlin 1959.

Hertwig, Richard, *Lehrbuch der Zoologie,* Jena 1916.

Hoenn, Karl, *Artemis,* Zurich 1946.

Kearton, Cherry, *The Island of Penguins,* Longman Green, London 1930.

Kerényi, Karl, 1951, *The Gods of the Greeks,* Thames and Hudson, London.

— , 1959, *The Heroes of the Greeks,* London.

Kleesattel, Walter, *Die Welt der lebenden Fossilien, Eine Reise in die Urzeit,* Darmstadt 2001.

Kluge, Friedrich, *Etymologisches Wörterbuch der deutschen Sprache,* Berlin 1960.

König, Bertha, *Meine Kindheits- und Lebenserinnerungen,* Karl König Archive.

König, Karl, *The Animals and their Destiny,* Camphill Books, Botton 2002.

—, *Kaspar Hauser and Karl König.* Floris Books, Edinburgh 2012.

—, *Die Ordnung der Tiere im Tierkreis,* 3 booklets (ed. Ernst Marti) Arlesheim, Switzerland.

Kraus, Eugen, *Das Aalproblem der modernen Biologie,* Goetheanum, Dornach 1932.

Lilly, John C. *Man and Dolphin,* Gollancz, London 1962.

Lindahl, *see* Curry-Lindahl, K.

Lockley, Ronald Mathias, *The Seals and the Curragh,* Dent, London 1954.

Lucanus, Friedrich von, *Zugvögel und Vogelzug,* Springer, Berlin 1929.

Makatsch, Wolfgang, *Die Vögel der Erde,* Berlin 1954.

Marret, Mario, *Sieben Mann bei den Pinguinen,* Kümmerley & Frey, Bern 1956.

Muir-Evans, Harold, *Sting-Fish and Seafarer,* Faber, London 1943.

Müller-Weidemann, Hans, *Karl König,* Camphill Books, 1996.

Oken, Lorenz, *Allgemeine Naturgeschichte für alle Stände,* Stuttgart 1837.

—, *Lehrbuch der Naturphilosophie,* Jena 1831.

Partridge, Eric, *Origins: A Short Etymological Dictionary.* 1966.

Payne, Katy, *Silent Thunder: The Hidden Voice of Elephants,* Weidenfield & Nicolson, London 1998.

Poppelbaum, Hermann, *Tierwesenskunde,* Dornach 1954.

Portmann, Adolf, *Die Tiergestalt,* Basle 1960.

—, *Von Vögeln und Insekten*, Rheinhadt, Basle 1957.

Pritzwald, F.P. Stegmann von, *Die Rassengeschichte der Wirtschaftstiere*, Jena 1924.

Rabinovitch, Melitta, *Der Delphin in Sage und Mythos der Griechen*, Hybernia, Dornach 1947.

Reinhardt, Ludwig, *Kulturgeschichte der Nutztiere*, Munich 1912.

Rivolier, Jean, *Emperor Penguins*, Elek, London 1956.

Roule, Louis, *Fishes, their Journeys and Migrations*, Routledge, London 1933.

Schadewaldt, Wolfgang, *Die Sternsagen der Griechen*, Fischer, Frankfurt.

Scheffer, Victor Blanchard, *Seals, Sea Lions and Walruses*, Stanford University Press, 1958.

Schulz, H. *Weissstorchzug, Ökologie, Gefährdung und Schutz des Weissstorches in Afrika und Nahost*, Weikersheim 1988.

Schwabe, Julius, *Archetyp und Tierkreis*, Basel 1951.

Selg, Peter, *Karl König: My Task*, Floris Books, Edinburgh 2008.

Siewert, Horst, *Störche*, Gütersloh 1955.

Slijpers, Everhard Johannes, *Whales*, Hutchinson, London 1962.

Steiner, Rudolf. Volume Nos refer to the Collected Works (CW), or to the German Gesamtausgabe (GA).

—, *Antworten der Geisteswissenschaft auf die grossen Fragen des Daseins*, GA 60, Dornach 1983.

—, *The Apocalypse of St John*, CW 104, Rudolf Steiner Press, UK 1977.

—, *The Boundaries of Natural Science*, CW 322, Anthroposophic Press, USA 1983.

—, *Christ and the Spiritual World*, CW 149, Rudolf Steiner Press, UK 2008.

—, *The Christ Impulse and the Development of the Ego Consciousness*, CW 116, Anthroposophic Press, USA 1976.

—, *Cosmic Memory, Prehistory of Earth and Man*, CW 11, Steinerbooks, USA 2006.

—, *The East in the Light of the West*, CW 113, (translated from *Der Orient im Lichte des Okzidents*) Anthroposophical Publishing Company, London 1940.

—, *Egyptian Myths and Mysteries*, CW 106, Anthroposophic Press, USA 1971.

—, *Die Erkenntnis der Seele und des Geistes*, GA 56, Dornach 1985.

—, *An Esoteric Cosmology*, CW 94, SteinerBooks, USA 2008.

—, *Founding a Science of the Spirit*, CW 95, Rudolf Steiner Press, UK 1999.

—, *From Comets to Cocaine: Answers to Questions*, CW 348, Rudolf Steiner Press, UK 2001.

—, *From Elephants to Einstein: Answers to Questions*, CW 352, Rudolf Steiner Press, UK 1998.

—, *Fundamentals of Esotericism*, CW 93a, Rudolf Steiner Press, UK 1983.

—, *Das Geheimnis der Trinität*, GA 214, Dornach 1989.

—, *Geisteswissenschaftliche Erläuterung zu Goethes Faust*, Vol. 2, GA 273, Dornach 1981.

—, *The Gospel of St John and its Relation to the Other Gospels,* CW 112, Anthroposophic Press, USA 1982.

—, *The Gospel of St Matthew,* CW 123, Rudolf Steiner Press, UK 1965.

—, *Harmony of the Creative Word,* CW 230, Rudolf Steiner Press, UK 2001.

—, *The Karma of Untruthfulness,* Vol, 1, CW 173, Rudolf Steiner Press, UK 2005.

—, *Macrocosm and Microcosm,* CW 119, Anthroposophic Press, USA 1985.

—, *Menschenwerden, Weltenseele und Weltengeist,* Vol. 1, GA 205, Dornach 1967.

—, *The Mission of Folk-Souls in Relation to Teutonic Mythology,* CW 121, Rudolf Steiner Press, UK 2005.

—, *The Mystery of the Trinity,* (part of GA 214, *Das Geheimnis der Trinität)* Anthroposophic Press, USA 1947.

—, *Occult History,* CW 126, Rudolf Steiner Press, UK 1982.

—, *Die okkulten Wahrheiten alter Mythen und Sagen,* GA 92, Dornach 1999.

—, *Der Orient im Lichte des Okzidents,* GA 113, Dornach 1982.

—. *The Redemption of Thinking: A Study in the Philosophy of Thomas Aquinas,* CW 74, Anthroposophic Press, USA 1983.

—, *The Riddles of Philosophy,* CW 18, Anthroposophic Press, USA 1973.

—, *Universe, Earth and Man,* CW 105, Rudolf Steiner Press, UK 1955.

—, *Wonders of the World, Ordeals of the Soul, Revelations of the Sprit,* CW 129, Rudolf Steiner Press, UK 1983.

Summers-Smith, D. *The House Sparrow,* Collins, London 1963.

Wachsmuth, Guenther, *Die Entwicklung der Erde,* Philosophisch-Anthroposophischer Verlag, Dornach 1950.

—, *The Evolution of Mankind,* Philosophisch-Anthroposophischer Verlag, Dornach 1961.

Wilckens, Martin, *Grundzüge der Naturgeschichte der Haustiere,* Leipzig 1905.

Young, J.Z. *The Life of Vertebrates,* Oxford 1950.

Index

Karl König's collected works are being published in English by Floris Books and in German by Verlag Freies Geistesleben. They encompass the entire, wide-ranging literary estate of Karl König, including his books, essays, manuscripts, lectures, diaries, notebooks, his extensive correspondence and his artistic works, across twelve subjects.

Karl König Archive subjects

Medicine and study of the human being
Curative education and social therapy
Psychology and education
Agriculture and science
Social questions
The Camphill movement
Christianity and the festivals
Anthroposophy
Spiritual development
History and biographies
Artistic and literary works
Karl König's biography

Karl König Archive
www.karlkoeniginstitute.org
office@karlkoeniginstitute.org

Printed in the USA
CPSIA information can be obtained
at www.ICGtesting.com
JSHW011520221024
72172JS00015B/125

9 780863 159664